Ready, Set, Go!

Geometry

Workbook

Mel Friedman, M.S.

For New Jersey HSPA

Research & Education Association
Visit our website at
www.rea.com

Research & Education Association
61 Ethel Road West
Piscataway, New Jersey 08854
E-mail: info@rea.com

REA's Ready, Set, Go!™
Geometry Workbook
for the New Jersey High School Proficiency Assessment (HSPA)

Printed in the United States of America

ISBN-13: 978-0-7386-0522-7
ISBN-10: 0-7386-0522-0

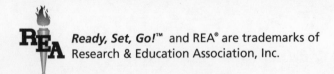

Ready, Set, Go!™ and REA® are trademarks of
Research & Education Association, Inc.

Contents

Welcome
to the *Ready, Set, Go! Geometry Workbook!*

About This Book

This book will help high school math students at all learning levels understand basic geometry. Students will develop the skills, confidence, and knowledge they need to succeed on high school math exams with emphasis on passing high school graduation exams.

More than 20 easy-to-follow lessons break down the material into the basics. In-depth, step-by-step examples and solutions reinforce student learning, while the "Math Flash" feature provides useful tips and strategies, including advice on common mistakes to avoid.

Students can take drills and quizzes to test themselves on the subject matter, then review any areas in which they need improvement or additional reinforcement. The book concludes with a final exam, designed to comprehensively test what students have learned.

The *Ready, Set, Go! Geometry Workbook* will help students master the basics of mathematics—and help them face their next math test—with confidence!

Icons Explained

Icons make navigating through the book easier. The icons, explained below, highlight tips and strategies, where to review a topic, and the drills found at the end of each lesson.

Look for the **"Math Flash"** feature for helpful tips and strategies, including advice on how to avoid common mistakes.

When you see the **"Let's Review"** icon, you know just where to look for more help with the topic on which you are currently working.

The **"Test Yourself!"** icon, found at the end of every lesson, signals a short drill that reviews the skills you have studied in that lesson.

To the Student

This workbook will help you master the fundamentals of Algebra & Functions. It offers you the support you need to boost your skills and helps you succeed in school and beyond!

It takes the guesswork out of math by explaining what you most need to know in a step-by-step format. When you apply what you learn from this workbook, you can

1. do better in class;
2. raise your grades, and
3. score higher on your high school math exams.

Each compact lesson in this book introduces a math concept and explains the method behind it in plain language. This is followed with lots of examples with fully worked-out solutions that take you through the key points of each problem.

The book gives you two tools to measure what you learn along the way:

✔ Short drills that follow <u>each</u> lesson
✔ Quizzes that test you on <u>multiple</u> lessons

These tools are designed to comfortably build your test-taking confidence.

Meanwhile, the "Math Flash" feature throughout the book offers helpful tips and strategies—including advice on how to avoid common mistakes.

When you complete the lessons, take the final exam at the end of the workbook to see how far you've come. If you still need to strengthen your grasp on any concept, you can always go back to the related lesson and review at your own pace.

To the Parent

For many students, math can be a challenge—but with the right tools and support, your child can master the basics of algebra. As educational publishers, our goal is to help all students develop the crucial math skills they'll need in school and beyond.

This *Ready, Set, Go! Workbook* is intended for students who need to build their basic geometry skills. It was specifically created and designed to assist students who need a boost in understanding and learning the math concepts that are most tested along the path to graduation. Through a series of easy-to-follow lessons, students are introduced to the essential mathematical ideas and methods, and then take short quizzes to test what they are learning.

Each lesson is devoted to a key mathematical building block. The concepts and methods are fully explained, then reinforced with examples and detailed solutions. Your child will be able to test what he or she has learned along the way, and then take a cumulative exam found at the end of the book.

Whether used in school with teachers, for home study, or with a tutor, the ***Ready, Set, Go! Workbook*** is a great support tool. It can help improve your child's math proficiency in a way that's fun and educational!

To the Teacher

As you know, not all students learn the same, or at the same pace. And most students require additional instruction, guidance, and support in order to do well academically.

Using the Curriculum Focal Points of the National Council of Teachers of Mathematics, this workbook was created to help students increase their math abilities and succeed on high school exams with special emphasis on high school proficiency exams. The book's easy-to-follow lessons offer a review of the basic material, supported by examples and detailed solutions that illustrate and reinforce what the students have learned.

To accommodate different pacing for students, we provide drills and quizzes throughout the book to enable students to mark their progress. This approach allows for the mastery of smaller chunks of material and provides a greater opportunity to build mathematical competence and confidence.

When we field-tested this series in the classroom, we made every effort to ensure that the book would accommodate the common need to build basic math skills as effectively and flexibly as possible. Therefore, this book can be used in conjunction with lesson plans, stand alone as a single teaching source, or be used in a group-learning environment. The practice quizzes and drills can be given in the classroom as part of the overall curriculum or used for independent study. A cumulative exam at the end of the workbook helps students (and their instructors) gauge their mastery of the subject matter.

We are confident that this workbook will help your students develop the necessary skills and build the confidence they need to succeed on high school math exams.

About Research & Education Association

Founded in 1959, Research & Education Association (REA) is dedicated to publishing the finest and most effective educational materials—including software, study guides, and test preps—for students in elementary school, middle school, high school, college, graduate school, and beyond.

Today, REA's wide-ranging catalog is a leading resource for teachers, students, and professionals.

We invite you to visit us at *www.rea.com* to find out how "REA is making the world smarter."

About the Author

Author Mel Friedman is a former classroom teacher and test-item writer for Educational Testing Service and ACT, Inc.

Acknowledgments

We would like to thank Larry Kling, Vice President, Editorial, for his editorial direction; Pam Weston, Vice President, Publishing, for setting the quality standards for production integrity and managing the publication to completion; Alice Leonard, Senior Editor, for project management and preflight editorial review; Diane Goldschmidt, Senior Editor, for post-production quality assurance; Ruth O'Toole, Production Editor, for proofreading; Rachel DiMatteo, Graphic Artist, for page design; Christine Saul, Senior Graphic Artist, for cover design; and Jeff LoBalbo, Senior Graphic Artist, for post-production file mapping

We also gratefully acknowledge Heather Brashear for copyediting, and Kathy Caratozzolo of Caragraphics for typesetting.

Thank you to author Bob Miller for his review of all test items.

Orsolina Cetta, Jamie Chaikin and the students of Piscataway High School, Piscataway, NJ, for reviewing and field-testing lessons from this book.

Basic Properties of Points, Rays, Lines, and Angles

In this lesson, we will look at the features and properties of points, lines, line segments, rays, and angles. Each lesson will use at least one of these elements, and many lessons will use all five of them. In real life, we see these elements everywhere. We find them, for example, as (a) the tip of a pen, (b) an arrow, (c) a piece of string, and (d) the path of a billiard ball in a game of pool.

Your Goal: When you have completed this lesson, you should be familiar with the properties of points, rays, lines, and angles. In addition, you will be able to correctly name each of them, using the correct letters.

LESSON 1

Basic Properties of Points, Rays, Lines, and Angles

All geometric shapes that you will meet in this workbook are considered to lie in a single plane. A **plane** is essentially a flat surface with length and width, but no depth. A plane extends indefinitely. A good way to picture a plane is to think of a "magic" carpet that extends in all directions but has no thickness.

The most basic unit in geometry is a **point**. Examples are pencil points, tiny seeds, and dots. A point has location, but no dimension; thus, it is infinitely small. Points are named by using a capital letter. Here are examples of point *P*, point *Q*, and point *R*.

• P   • Q   • R

MathFlash!

Technically, these drawn points have some length and width (although very small). However, they are only <u>visual representations</u> of points.

A **line** is a row of points that can all be connected with a ruler. Lines are named by using any two of its points, written in capital letters, with the symbol ↔ over the letters that name or identify the two points. The letters used do not have to be in alphabetical order. Here are examples, with notation, of line *AB*, line *DC*, and line *EH*:

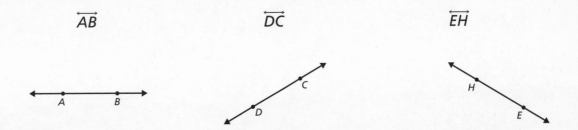

A **line segment** is a finite portion of a line. "Finite" means it has a beginning and an end. It has two endpoints that are named with capital letters, with the bar symbol over the two letters. As with lines, the letters of the points used do not have to be in alphabetical order. Here are examples, with notation, of line segments *JK*, *NP*, and *XY*:

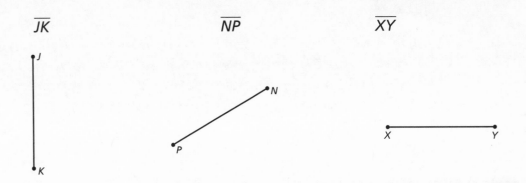

MathFlash!

Lines and line segments may face in any direction, but the notation over the associated letters is always horizontal. Both lines and line segments have length, but no width. Line segments have measurable length, whereas the length of a line cannot be measured.

A **ray** is basically a half line in that it consists of a starting point, called its **end point**, and goes forever in the direction of a line. It is named by using its end point first, followed by any other point that it contains. The symbol → is placed over the two capital letters that are used to denote the ray. Rays have length, but no width. Here are examples, with notation, of rays *MQ*, *SR*, and *VT*:

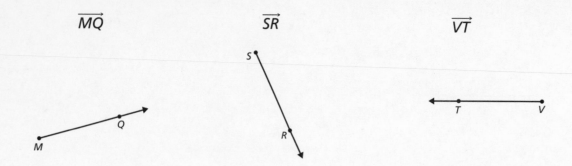

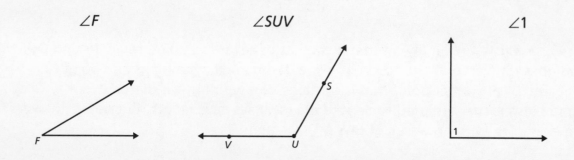

An **angle** consists of two rays that share a common end point. The only restriction is that one of the rays must not overlap the other. The common end point is called the **vertex**. (The plural of "vertex" is "vertices.") The two rays that form the angle are called **sides**. Using a protractor, an angle can be measured by the "opening" between its rays. The unit of measurement is commonly called a **degree**. An angle is named in one of three ways:

      (a) by its vertex;

      (b) by a point on one ray, followed by its vertex and then a point
         on the other ray; and

      (c) by a number.

The symbol $\angle$ is placed before the letter(s) used to name an angle. Here are examples, with notation, of angle *F*, angle *SUV*, and angle 1:

MathFlash!

The symbol ∠ is used even if the angle appears larger. In addition, an angle may actually be a line, since the two rays do not overlap. Numbers used to name angles are generally low integers. So if a diagram had three angles next to each other, they would usually be named ∠1, ∠2, and ∠3. Here is the associated diagram.

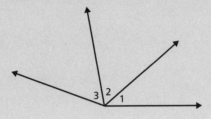

1 **Example:** *How can the following line be named in three different ways?*

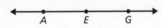

Solution: Using any two letters in any order, we could write $\overleftrightarrow{AG}$, $\overleftrightarrow{EG}$, or $\overleftrightarrow{EA}$.

2 **Example:** *In how many different ways can the following line segment be named?*

Solution: There are only two ways, namely, $\overline{HK}$ or $\overline{KH}$.

3 **Example:** *In how many different ways can the following angle be named?*

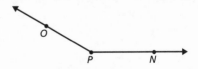

Solution: There are three ways, namely, ∠P, ∠OPN, and ∠NPO.

4 | **Example:**   *In the following diagram, in how many other ways can ∠1 be named?*

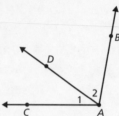

Solution: There are two ways, namely, ∠CAD and ∠DAC.

In Example 4, you cannot use ∠A to name ∠1. In fact, the symbol ∠A cannot be used for any of the angles shown, since there would be confusion as to which angle you are talking about. There are actually three distinct angles with a vertex at A. Also, ∠1 and ∠2 are called adjacent angles, since they share the same vertex and a common ray.

5 | **Example:**   *In the following diagram, which angle is adjacent to ∠FGK?*

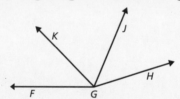

Solution: ∠KGJ is adjacent to ∠FGK, since they share the vertex G and the common ray $\overrightarrow{GK}$.

6 | **Example:**   *In the following diagram, how do you name the rays that form ∠3?*

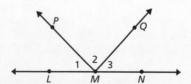

Solution: The two rays forming ∠3 are $\overrightarrow{MQ}$ and $\overrightarrow{MN}$.

7 **Example:**   *In the following diagram, which angles may be named using only the vertex?*

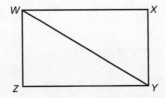

Solution: Only ∠X and ∠Z can be named using only the vertex. At each of points W and Y there are actually three angles.

8 **Example:**   *Returning to Example 7, how do you name the three angles at Y?*

Solution: The largest angle at Y is ∠ZYX. The two smaller angles are ∠ZYW and ∠XYW.

9 **Example:**   *How do you name the following ray in three different ways?*

Solution: The three ways are $\overrightarrow{XB}$, $\overrightarrow{XA}$, and $\overrightarrow{XC}$.

10 **Example:**   *How do you name the following angle in six different ways?*

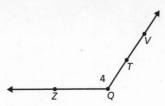

Solution: The six ways are ∠4, ∠Q, ∠ZQT, ∠ZQV, ∠TQZ, and ∠VQZ.

Test Yourself!

1. **How many rays are needed to define an angle?**

    Answer: _____

2. **How many points does a line have?**

    Answer: _____

3. **How do you name the following angle in three ways?**

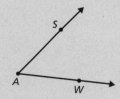

    Answers: _____

    _____

4. **How do you name the angle that is adjacent to ∠YSN?**

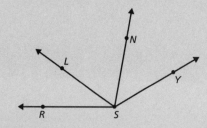

    Answer: _____

    _____

 Test Yourself! (continued)

5. In the following diagram, how do you name the three different line segments?

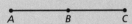

A   B   C

Answers: _____

6. Which one of the following would most closely resemble a ray?

    (A) Paper clip   (C) Stop sign

    (B) One-way street sign   (D) Long wire

7. Which one of the following correctly describes a line segment?

    (A) Length and width   (C) Neither length nor width

    (B) Width, but no length   (D) Length, but no width

8. Look at the following figure.

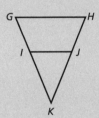

Which angle(s) can be correctly described using only the vertex letter?

    (A) $\angle G$, $\angle H$, and $\angle K$ (C) $\angle I$, and $\angle J$

    (B) $\angle G$, $\angle H$, $\angle I$, $\angle J$, and $\angle K$ (D) None of the angles

9.   **How many letters are needed to identify a point?**

Answer: _____

10.   **How do you name the following angle in five different ways?**

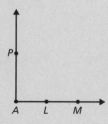

Answers: _____

MathFlash!

An angle is defined as the formation of two rays. But, within a geometric figure such as the one that appears in question #8, angles are shown as the formation of two line segments. The <u>entire</u> ∠K would be seen from this diagram.

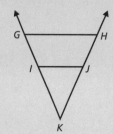

Measuring Line Segments and Angles

In this lesson, we will explore how to measure line segments and angles using their appropriate units. We cannot measure points, lines, or rays. Remember that points have no dimension, and both rays and lines extend indefinitely. A line segment has a specific length, and an angle has a specific "opening."

Your Goal: When you have completed this lesson, you should be familiar with the ways in which line segments and angles are measured.

LESSON 2

Measuring Line Segments and Angles

Given a line segment, $\overline{AB}$, we can simply calculate, using a ruler, the number of inches between points A and B. This is called the distance of $\overline{AB}$ and is denoted as AB or BA.

In most calculation problems concerning line segments, line segments will have numbers artificially assigned to them. This means that if you see a line segment that has the number 4 associated with it, you would say that its length is 4 units, where a **unit** is some specific (but arbitrary) length. This assigned number is called the **measure of the line segment**.

To clarify this concept, you may see $\underset{C}{\bullet}\overset{10}{\rule{2cm}{0.4pt}}\underset{D}{\bullet}$ in one problem, but then see $\underset{E}{\bullet}\overset{4}{\rule{3.5cm}{0.4pt}}\underset{F}{\bullet}$ in another problem. Don't be disturbed that $\overline{CD}$ has a higher number than $\overline{EF}$. The length of the unit used in each case is different. However, within a single problem or single diagram, the length of the unit used <u>must</u> be consistent. Thus, if $\overline{GH}$ appears as $\underset{G}{\bullet}\rule{1.5cm}{0.4pt}\underset{H}{\bullet}$ and has a measure of 5, then if $\overline{IJ}$ appears as $\underset{I}{\bullet}\rule{3cm}{0.4pt}\underset{J}{\bullet}$ in the same problem, it must have a measure higher than 5.

Often, line segments (as well as other geometric figures) are not drawn to scale. However, the figures must be drawn in a reasonable manner with respect to their lengths. Can you imagine how much paper (and wasted effort!) would be needed for any book if the actual dimensions of most objects were used?

Angle measurement is more standard. A protractor is used to measure the "opening" of any angle. The most common unit of measure is the **degree**, indicated by the symbol °. This symbol is placed in the upper right corner of the number associated with the angle. For example, if an angle were 40 degrees, it would be written as 40°. To measure an angle, place the center mark of the protractor on the vertex. Then align the zero degree measure with one side of the angle. The measure of the angle will be the number on the protractor that corresponds to the other side of the angle.

Notice that the protractor has two rows of numbers. If the angle opening is measured in a clockwise rotation, use the lower row of numbers on the protractor; if it is measured in a counterclockwise rotation, use the upper layer. Although the physical size of an angle can be as large as 360°, the largest angle we will use in most lessons is 180°. When referring to the **measure of an angle**, the notation to be used is *m∠*_____ = _____°. For example, if angle *ABC* has a measure of 30 degrees, we would write *m∠ABC* = 30°.

Here are some examples of angles with their associated measures. In each case, the protractor is placed on top of the angle so that you can see the measurement.

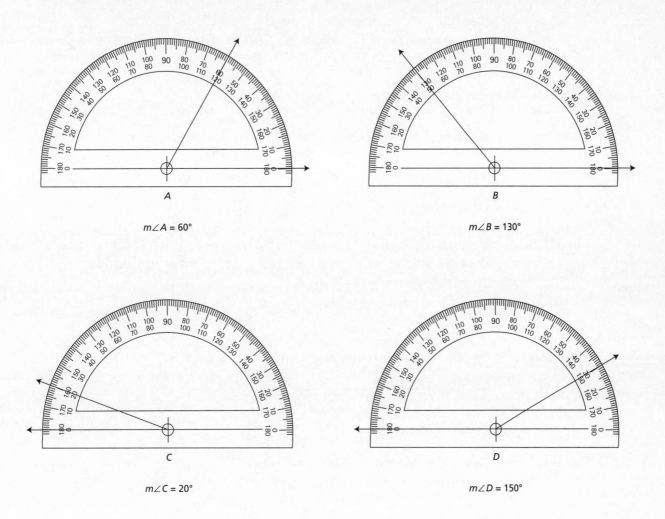

m∠A = 60°

m∠B = 130°

m∠C = 20°

m∠D = 150°

Here is a drawing for $\angle PQR$ if $m\angle PQR = 180°$.

Basically, an angle whose degree measure is 180° is actually a line. If $m\angle XYZ = 0°$, the drawing could appear as follows:

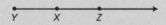

In almost all problems you will see, a protractor is not given to check the measure of the angles.

1 **Example:**   *For the following line segment, AC = 6 and CE = 5.*

What is the value of AE?

Solution: Since all three points lie on the same line segment, we can see that $AC + CE = AE$. This is called the **Segment Addition Rule**. Thus, AE is simply $6 + 5 = 11$.

2 **Example:**   *For the following line segment, BD = 4, DF = 13, and BH = 19.*

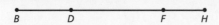

What is the value of FH?

Solution: Using the Segment Addition Rule, $BD + DF + FH = BH$. Let x stand for the value of FH. Then, substituting the numbers, we get $4 + 13 + x = 19$. Hopefully, your algebra skills are up to speed! This equation simplifies to $17 + x = 19$, so $x = 2$.

3 **Example:** *For the following line segment,* $\overline{MN}$ *is twice as large as* $\overline{LM}$ *and LN = 42.*

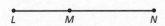

What is the value of LM?

Solution: Let x represent the length of $\overline{LM}$, and let $2x$ stand for the length of $\overline{MN}$. Since $LM + MN = LN$, when we put in the numbers we get $x + 2x = 42$. This equation simplifies to $3x = 42$, so $x = 14$. This means $LM = 14$.

MathFlash!

If you were asked to determine MN, just calculate (2)(14) = 28.

4 **Example:** *For the following line segment,* $\overline{FL}$, *is six units longer than* $\overline{LG}$, *and FG = 63.*

What is the value of FL?

Solution: Let x represent the length of $\overline{LG}$, and let $x + 6$ represent the length of $\overline{FL}$. $FL + LG = FG$, so we can write $(x + 6) + x = 60$. This equation becomes $2x + 6 = 60$. Subtracting 6 from each side, we get $2x = 54$. Then $x = 27$.
Don't stop here! Remember that the length of $\overline{FL}$ is represented by $x + 6$. The value of FL is $27 + 6 = 33$.

5 **Example:**   *For the following line segment, PR = 10, QR = 3, and QS = 16.*

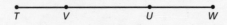

P   Q   R   S

What is the value of PS?

Solution: Be careful! You cannot simply add these three numbers, since $\overline{PR}$ and $\overline{QS}$ overlap. They both contain $\overline{QR}$.
Suppose we were to add the values of *PR* and *QS*. Their sum is 26, which is not correct because you have added *QR* twice.
So all you would need to do is subtract the value of *QR* from 26; this would give you the correct answer of 23.

6 **Example:**   *For the following line segment, TV = UW, UV = 15, and TW = 34.*

T   V   U   W

What is the value of TV?

Solution: Let *x* represent the value of each of *TV* and *UW*. By the Segment Addition Rule, *TV* + *UV* + *UW* = *TW*. By substitution, we get *x* + 15 + *x* = 34. From this equation comes the following sequence of steps: 2*x* + 15 = 34, 2*x* = 19, and finally *x* = 9.5.

7 **Example:**   *For the following diagram, m∠1 = 42° and m∠ABC = 106°.*

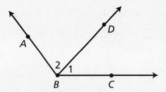

What is the value of m∠2?

Solution: Since ∠1 and ∠2 are adjacent angles, we can see that *m*∠1 + *m*∠2 = ∠*ABC*. This is called the **Angle Addition Rule**. Looking back to Example 1, this rule is very similar to the Segment Addition Rule. The Angle Addition Rule assures us that when two angles are adjacent, the measure of the composite angle that is formed is simply the sum of the measures of the two adjacent angles.
Let *x* represent the measure of ∠2. Then 42° + *x* = 106°, so *x* = 64°.

8 **Example:** *For the following diagram, the measure of ∠EFG is 16° greater than the measure of ∠HFG and m∠EFH = 84°.*

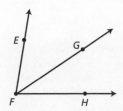

What is the value of m∠EFG?

Solution: As in Example 7, we can use the Angle Addition Rule.
Let x represent the measure of ∠HFG, and let $(x + 16)$ represent the measure of ∠EFG.
Then $x + (x + 16) = 84°$.
Simplifying, we get $2x + 16 = 84°$.
This leads to $2x = 68$, so that $x = 34°$.
However, be sure you don't stop here.
The value of 34° refers to the measure of ∠HFG.
To find ∠EFG, add 16° to 34° to get the correct answer of 50°.

9 **Example:** *For the following diagram, m∠1 = 20°, m∠IJK = 128°, and m∠3 = 81°*

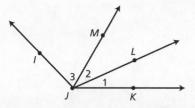

What is the value of m∠2?

Solution: Similar to Example 8, we will use an extension of the Angle Addition Rule, namely, that $m∠1 + m∠2 + m∠3 = m∠IJK$.
Let x represent $m∠2$.
Then by substitution, $20 + x + 81 = 128$.
$101 + x = 128$.
Thus, $x = 27°$.

10 **Example:**   *For the following diagram, m∠NPR = 45°, m∠SPQ = 35°, and m∠2 = 24°.*

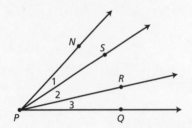

What is the measure of m∠NPQ?

Solution: This is similar to the one we worked on in Example 5. See that
m∠1 + m∠2 + m∠3 = m∠NPQ.
Now add the measures of m∠NPR and m∠SPQ to get 80°.
We have actually added (by substitution) (m∠1 + m∠2) +
(m∠2 + m∠3). This sum includes m∠2 <u>twice</u>.
All we need to do is to subtract the value of m∠2 from this sum.
We will then have m∠1 + m∠2 + m∠3, which is the value of
m∠NPQ.
Thus, we can see that m∠NPQ = 80 − 24 = 56°.

MathFlash!

Another way to solve Example 10 would have been to figure out the value of each of m∠1 and m∠3. Then simply add these numbers to 24°, which is the value of m∠2.
Since m∠NPR = 45° and m∠2 = 24°, by subtraction m∠1 = 21°.
Likewise, since m∠SPQ = 35° and m∠2 = 24°, by subtraction m∠3 = 11°. Finally, m∠NPQ = 21 + 24 + 11 = 56°.

Test Yourself!

1. Suppose you are given that *MN* = 12 units in one problem. In a different problem, you are given that *PQ* = 6 units. Which one of the following is <u>always</u> true?

    (A) $\overline{PQ}$ must be shorter than but not necessarily half the size of $\overline{MN}$.

    (B) $\overline{PQ}$ must be half the size of $\overline{MN}$.

    (C)   The length of a unit is the same in each problem.

    (D)   It is impossible to determine which segment is larger.

2. Think about the following geometric items: lines, line segments, rays, and angles. Which of these <u>cannot</u> be measured?

    (A)   Lines and rays

    (B)   Line segments and angles

    (C)   Lines and angles

    (D)   Line segments and rays

3. Using your protractor, which one of the following angles has a measure of 55°?

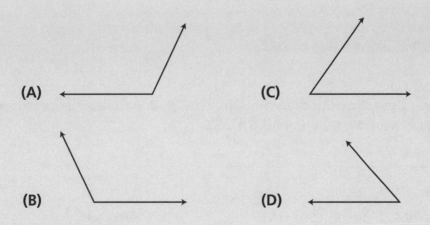

Test Yourself! (continued)

4. Using your protractor, which one of the following angles has a measure greater than 110º but less than 130º?

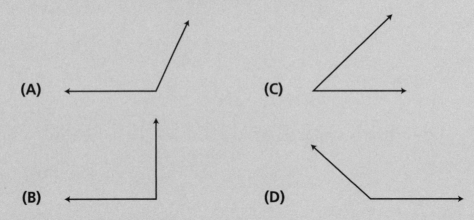

(A)

(C)

(B)

(D)

5. For the following diagram, points *C, J,* and *M* all lie on the same line. *CJ* = 13, and *CM* = 21.

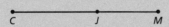

What is the value of *MJ* ?   Answer:_____

6. For the following diagram, points *D, F,* and *H* all lie on the same line. *DH* = 85, and $\overline{DF}$ is eleven units larger than $\overline{FH}$.

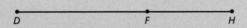

What is the value of *DF*?   Answer:_____

7. For the following diagram, points *G, K, L,* and *N* all lie on the same line. *KL* = 19, *NL* = 6, and *GN* = 39.

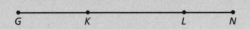

What is the value of *GK*?   Answer:_____

8. For the following diagram, points *Q, R, S,* and *T* all lie on the same line. *QS* = 24, *RT* = 33, and *RS* = 16.

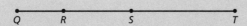

What is the value of *TQ*?   Answer:_____

9. For the following diagram, the measure of *m∠UVX* is 62° larger than the measure of *m∠XVW*. Also, *m∠UVW* = 138°.

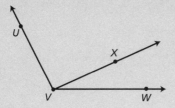

What is the value of *m∠UVX*?   Answer:_____

10. For the following diagram, *m∠CXZ* = 17°, *m∠BXY* = 169°, and the measure of *∠BXC* is seven times as large as the measure of *∠ZXY*.

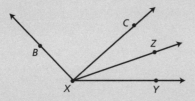

What is the value of *m∠BXC*?   Answer:_____

11.   In which one of the following situations would you be absolutely certain that $\angle ABC$ and $\angle CBD$ are adjacent angles?

(A) $m\angle ABC = 23°$, $m\angle CBD = 28°$, and $m\angle ABD = 33°$

(B) $m\angle ABC = 8°$, $m\angle CBD = 16°$, and $m\angle ABD = 12°$

(C) $m\angle ABC = m\angle CBD$

(D) $m\angle ABC = 51°$, $m\angle CBD = 71°$, and $m\angle ABD = 122°$

12.   For the following diagram, $m\angle EFH = 90°$, $m\angle JFH = 76°$, and $m\angle EFG = 132°$.

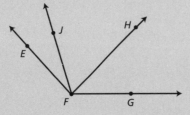

What is the value of $m\angle JFG$?   Answer:_____

Categories of Angles and Lines—Part I

In this lesson, we will explore the basic ways to name angles and lines. The names of the angles are directly related to their sizes. While the names associated with lines are related to the positions of any two of them.

Your Goal: When you have completed this lesson, you should be familiar with the name given to a specific size angle and the label associated with any given pair of lines.

LESSON 3

Categories of Angles and Lines— Part 1

From the information you learned in Lesson 2, recall that the size of an angle lies between 0° and 180°, inclusive. Any angle whose measure is less than 90° is called an **acute** angle. (No, the other angles are not called ugly!) Below are two examples, with their actual measures.

1 **Example:**   **2** **Example:**

50°   25°

If an angle measures exactly 90°, it is called a **right** angle. Below are two examples. Notice that in the second example, a small square is placed at the vertex. This is an accepted symbol to indicate that an angle has a measure of 90°.

3 **Example:**   **4** **Example:**

90°

If an angle's measure is greater than 90° but less than 180°, it is called an **obtuse** angle. Below are two examples, with their actual measures.

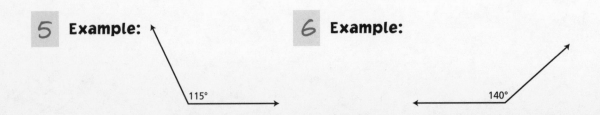

5 **Example:**   **6** **Example:**

115°   140°

If an angle measures exactly 180°, it is called a **straight** angle. Its appearance is actually a straight line. Below are two, examples identified as $m\angle ACD$ and $m\angle UZY$.

7 **Example:** ←•——•————•→
                 A   C   D

8 **Example:** ←•——•————•→
                 Y   Z   U

MathFlash!

In problems where you need to solve for the measure of an angle, the diagrams provided will not normally be drawn to scale. However, if an angle is described as being acute, then it must at least appear to be less than 90°. Also, if you solve for the measure of $\angle A$ in a given diagram and find that it is 165°, then $\angle A$ must appear to be obtuse.

If the sum of the measures of two angles is exactly 90°, they are called **complementary** angles. Below are two examples in which $\angle 1$ and $\angle 2$ are complementary. Notice that in the second example, the diagram has a small square at the vertex. In addition, the angles are adjacent, so their measures automatically add up to 90°.

9 **Example:**

10 **Example:**

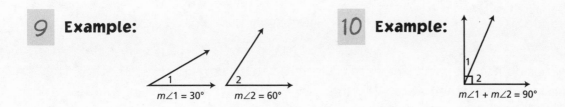

$m\angle 1 = 30°$ $m\angle 2 = 60°$ $m\angle 1 + m\angle 2 = 90°$

If the sum of the measures of two angles is exactly 180°, they are called **supplementary** angles. Below are two examples in which $\angle 3$ and $\angle 4$ are supplementary. Notice that in the second example, the diagram shows a pair of adjacent angles that form a line. This implies that their measures automatically add up to 180°.

11 **Example:**   12 **Example:**

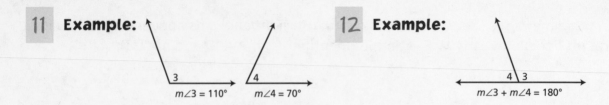

If you were to draw any two lines, call them $\overleftrightarrow{AB}$ and $\overleftrightarrow{CD}$, the relationship between them can appear in basically two different ways. Either they intersect at a point or they do not; these two possibilities are shown below.

13 **Example:**   14 **Example:**

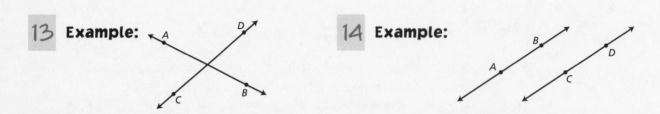

If they do intersect in such a way that the four angles formed each have a measure of 90°, the lines are considered to be **perpendicular** to each other. This relationship is shown mathematically by the symbol ⊥. Below are two examples in which $\overleftrightarrow{EF}$ is perpendicular to $\overleftrightarrow{GH}$, which is expressed mathematically as $\overleftrightarrow{EF} \perp \overleftrightarrow{GH}$. We could also write $\overleftrightarrow{GH} \perp \overleftrightarrow{EF}$.

Notice that a small square is placed in one of the four corners where the lines intersect.

15 **Example:**   16 **Example:**

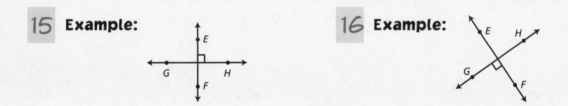

If the two lines do not intersect, they are considered to be **parallel** to each other. This relationship is shown mathematically by the symbol ‖. Below are two examples in which $\overleftrightarrow{IJ}$ is parallel to $\overleftrightarrow{KL}$, which is expressed mathematically as $\overleftrightarrow{IJ} \parallel \overleftrightarrow{KL}$.
We could also write $\overleftrightarrow{KL} \parallel \overleftrightarrow{IJ}$.

17 **Example:**   18 **Example:**

Now suppose that we look at two line segments, or even one line and one line segment.

In Example 19, $\overline{PQ}$ intersects $\overleftrightarrow{RS}$.   In Example 20, $\overline{PQ}$ intersects $\overline{RS}$.

19 **Example:** 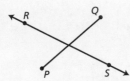   20 **Example:**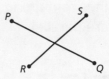

In Example 21, $\overline{TU}$ is parallel to $\overleftrightarrow{VX}$.   In Example 22, $\overline{TU}$ is parallel to $\overline{VX}$.

21 **Example:** 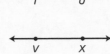   22 **Example:**

Of course, it is certainly possible that two line segments are perpendicular to each other or that a line segment is perpendicular to a line. There is also another situation that can occur between two line segments or between a line and line segment. They may not intersect or be parallel to each other. Look at the following two examples.

In Example 23, $\overline{MP}$ neither intersects nor is it parallel to $\overleftrightarrow{ST}$.   In Example 24, $\overline{MP}$ neither intersects nor is it parallel to $\overline{ST}$.

23 **Example:** 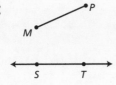   24 **Example:**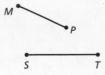

So we see that two lines must either intersect or be parallel to each other. If you are given either two line segments or one line and one line segment, then the three possibilities are that

  (a) they intersect,
  (b) they are parallel, or
  (c) they do not intersect but are not parallel.

Two line segments are parallel if the lines to which they belong are parallel. However, two line segments are perpendicular to each other only if they actually intersect and form 90° angles.

Look at the following two examples.

In Example 25, $\overline{AC}$ is parallel to $\overline{EG}$, since the lines containing these line segments are parallel.

In Example 26, $\overline{AC}$ is neither parallel to nor does it intersect $\overline{EG}$. Note that $\overline{AC}$ is not perpendicular to $\overline{EG}$, since they do not meet.

25 Example:.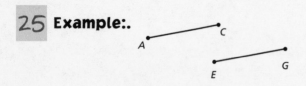

26 Example: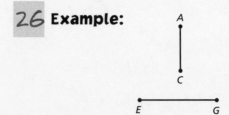

Be careful here! Even though it appears that $\overline{AC}$ would meet $\overline{EG}$ at a 90° angle by extending $\overline{AC}$, the two segments do not actually intersect.

As you probably realize, even a line segment and a line may be neither parallel to each other nor intersect each other, as shown in Examples 27 and 28. Once again, it may look as if $\overline{LM}$ would intersect $\overleftrightarrow{QR}$ by extending $\overline{LM}$ in Examples 27 and 28. However, they do not actually intersect because **a line segment doesn't go past its endpoints.**

27 Example:

28 Example:

1. Which one of the following could be the measure of an obtuse angle?

(A) 24°

(B) 75°

(C) 90°

(D) 132°

2. In which one of the following diagrams could you conclude that ∠1 and ∠2 are complementary angles?

(A)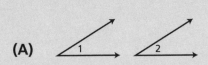

(C)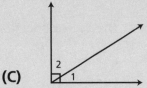

(B)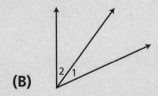

(D)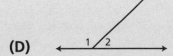

3. Suppose you add the measures of two obtuse angles. What conclusion must be true about their sum?

(A) It is exactly 90°.

(B) It is greater than 180°.

(C) It is between 90° and 180°.

(D) It is less than 90°.

4. Which one of the following pairs of angle measures would represent two supplementary angles?

(A) 36° and 144°

(B) 80° and 70°

(C) 13° and 77°

(D) 180° and 180°

Test Yourself! (continued)

5. $\overrightarrow{AD} \perp \overleftrightarrow{BF}$ describes which one of the following situations?

    (A)   Two line segments that are perpendicular

    (B)   A line segment that is perpendicular to a line

    (C)   Two line segments that are parallel

    (D)   A line segment that is parallel to a line

6.   Look at the following diagram.

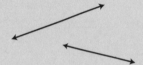

     Which one of the following <u>best</u> describes this diagram?

    (A)   Two line segments that never intersect

    (B)   Two lines that intersect

    (C)   Two line segments that are parallel

    (D)   Two lines that are parallel

7.   Which one of the following symbols is used to show that two lines are parallel?

    (A)   ∥   (C)   ∠

    (B)   ⊥   (D)   □

Test Yourself! (continued)

8. **Look at the following diagram.**

Which one of the following angles is shown to have a measure of 180°?

(A)   ∠CAB

(C)   ∠CTB

(B)   ∠TCA

(D)   ∠BAT

9. **Which one of the following would be an example of an angle that is neither acute, right, or obtuse?**

(A)   2°

(C)   190°

(B)   89°

(D)   101°

10. **In which one of the following diagrams is the measure of ∠TVX greater than 70° but less than 100°?**

(A)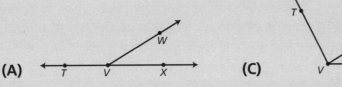

(C)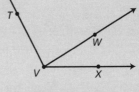

(B)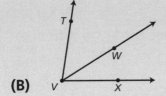

(D)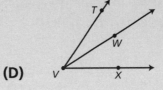

Categories of Angles and Lines—Part 2

In this lesson, we will continue to explore the basic categories of angles and lines. As with Lesson 3, angles are identified by their individual sizes, while lines and line segments are identified by their positions. In this lesson, we will use notation such as ℓ_1 and ℓ_2 to represent lines. Line segments will still be named by the use of their end points.

Your Goal: When you have completed this lesson, you should be familiar with the labels and relationships of angles and lines as they are used with parallel and intersecting lines.

LESSON 4

Categories of Angles and Lines— Part 2

We begin with a set of parallel lines and a third line that intersects both of them. Here is the corresponding diagram.

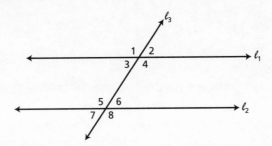

In this diagram, lines ℓ_1 and ℓ_2 are parallel to each other. Line ℓ_3 intersects both ℓ_1 and ℓ_2. The eight angles formed by the intersection of ℓ_3 with both ℓ_1 and ℓ_2 are shown as numbered angles.

First look at the angles formed by the intersection of ℓ_1 and ℓ_3. In particular, we call ∠1 and ∠4 a pair of **vertical** angles. Another name for vertical angles is **opposite** angles. This means that they are the nonadjacent angles for two intersecting lines. Similarly, ∠2 and ∠3 are also a pair of vertical angles. Can you spot a relationship between ∠1 and ∠4? If you sense that their measures are equal, you would be absolutely correct. When two lines intersect, each pair of vertical angles has equal measures. In the diagram, you can check that the measure of each of ∠1 and ∠4 is 125° and each of ∠2 and ∠3 is 55°.

Now look at the angles formed by the intersection of ℓ_2 and ℓ_3. Since ∠5 and ∠8 are a pair of vertical angles, their measures should be equal. Likewise, since ∠6 and ∠7 are a pair of vertical angles, their measures should also be equal. Let's go an extra step. Use your protractor to find that the measure of ∠5 is 125°, which matches the measures of ∠1, ∠4, and ∠8. Furthermore, if you check ∠6, you will find its measure to be 55°. This is no surprise since you already know that ∠5 and ∠6 must be supplementary angles.

Thus, the measures of ∠2, ∠3, ∠6, and ∠7 are equal to each other. The following examples will use a similar diagram, where the angle numbers are in the same location, but the measures are different.

1   **Example:**   *Suppose you are given that m∠1 = 108°. Which of the remaining seven angles have measures of 72°?*

   Solution: Each of ∠2, ∠3, ∠6, and ∠7 has a measure of 72°.

2   **Example:**   *Suppose you are given that the measure of ∠6 is 50° smaller than the measure of ∠4. What is the value of m∠4?*

   Solution: Let x represent the measure of ∠4, so that $x - 50$ represents the measure of ∠6. We recognize that ∠6 is supplementary to each of ∠1, ∠4, ∠5, and ∠8. Then $x + (x - 50) = 180°$. The next steps to solve this equation are $2x - 50 = 180°$, $2x = 230°$, and, finally, $x = 115°$.

3   **Example:**   *Suppose you are given that m∠8 is five times as large as m∠2. What is the value of m∠8?*

   Solution: Let x represent $m∠2$ and let $5x$ represent $m∠8$. By referring to the paragraph above Example 1, we can recognize that ∠2 and ∠8 are supplementary angles. Then $x + 5x = 180°$. So, $6x = 180°$, which leads to $x = 30°$. We need the value of $m∠8$, which is $(5)(30°) = 150°$.

As you look carefully at the diagram we have been using, notice that ∠1 and ∠5 are in the same relative position on lines ℓ_1 and ℓ_2. These are called **corresponding** angles and must have the same measure if ℓ_1 and ℓ_2 are parallel to each other. In the same way, ∠4 and ∠8 are corresponding angles. Based on this information, you should be able to identify the remaining two pairs of corresponding angles. Now notice that ∠3 and ∠6 lie in the "interior" of ℓ_1 and ℓ_2, respectively, and also on the opposite side of ℓ_3. They are called **alternate interior** angles. The other pair of alternate interior angles are ∠4 and ∠5. Alternate interior angles of parallel lines must have equal measures.

Can you guess the name for the pair of angles ∠1 and ∠8? The answer is **alternate exterior** angles. These angles lie in the exterior of ℓ_1 and ℓ_2, and on the opposite side of ℓ_3. The other pair of alternate exterior angles are ∠2 and ∠7. Alternate exterior angles of parallel lines must have equal measures. A line such as ℓ_3 that intersects two parallel lines is called a **transversal**.

If two parallel line segments and a transversal are given, the same definitions and equalities would apply. For example, corresponding angles would have the same measure. Of course, if a line segment were parallel to a line and you were given a transversal, all the previous comments made about two parallel lines would still apply.

Look at the following diagram.

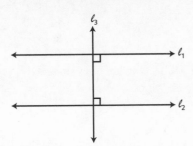

Line ℓ_3 is perpendicular to both ℓ_1 and ℓ_2. In this case, all eight angles formed by ℓ_3 with ℓ_1 and ℓ_2 have a measure of 90°. Notice that only a single small square is required at each intersection point.

Also, the statements about corresponding angles, alternate interior angles, and alternate exterior angles are reversible. For example, given ℓ_1 and ℓ_2 (with transversal ℓ_3), if the measures of a pair of alternate exterior angles are equal, then ℓ_1 and ℓ_2 must be parallel lines. The same conclusion would be reached if either a pair of corresponding angles or a pair of alternate interior angles had equal measures.

Let's introduce a few more concepts related to angles and lines. An **angle bisector** is a ray that divides a given angle into two angles of equal measure. Consider the following diagrams:

4 **Example:**

5 **Example:**

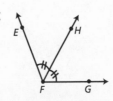

In Example 4, $\overrightarrow{BC}$ is the angle bisector of $\angle ABD$, so that $m\angle ABC = m\angle CBD$.
In Example 5, $\overrightarrow{FH}$ is the angle bisector of $\angle EFG$, so that $m\angle EFH = m\angle HFG$.
See the markings at points B and F.
The arc with the same number of tick marks shows that the angles have the same measure.

Another important concept with lines and line segments is a **perpendicular bisector**. A line ℓ_1 is the perpendicular bisector of $\overline{AB}$ if (a) ℓ_1 is perpendicular to $\overline{AB}$ and (b) ℓ_1 intersects $\overline{AB}$ at its midpoint. See Example 6. The midpoint of a line segment is simply the point that divides the given segment into two equal smaller segments. Note the single tick mark for each of $\overline{AC}$ and $\overline{CB}$; this means that $AC = CB$.

6 **Example:**

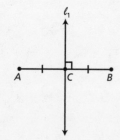

Example 7 illustrates one line segment as the perpendicular bisector of a second line segment. In this example, $\overline{IJ}$ is the perpendicular bisector of $\overline{KM}$. Note, however, that $\overline{KM}$ is <u>not</u> the perpendicular bisector of $\overline{IJ}$. As in Example 6, a small square is placed at the point of intersection to indicate that a 90° angle is formed. Also, a single tick mark is shown for each of $\overline{KN}$ and $\overline{NM}$.

7 **Example:**

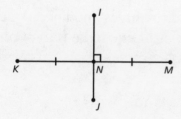

It is possible that two line segments are the perpendicular bisectors of each other. Example 8 shows $\overline{PU}$ and $\overline{RS}$ as the perpendicular bisectors of each other. Note that $\overline{PT}$ and $\overline{TU}$ have the same number of tick marks. The same is true for $\overline{RT}$ and $\overline{TS}$.

8 **Example:**

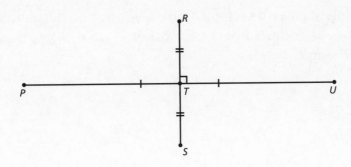

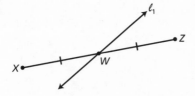

MathFlash!

The choice of which two segments receive double tick marks is purely arbitrary. Never assume that a segment with double tick marks is longer than one with a single tick mark.

It is possible that a line (or line segment) bisects a specific line segment but is <u>not</u> perpendicular to it. In Example 9, line ℓ_1 bisects $\overline{XZ}$ at point W, but they do not intersect at a 90° angle. In Example 10, $\overline{UY}$ bisects $\overline{DE}$ at point F, but these segments are not perpendicular to each other. Note that $\overline{DE}$ does <u>not</u> bisect $\overline{UY}$.

9 **Example:**

10   **Example:**

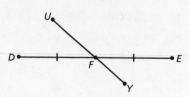

Finally, two line segments may bisect each other, yet not be perpendicular to each other. In Example 11, $\overline{AB}$ and $\overline{GH}$ bisect each other at point *J*, yet these two segments do not meet at a 90° angle.

11   **Example:**

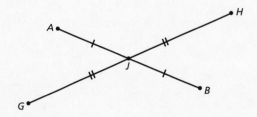

MathFlash!

Given two lines, ℓ_1 and ℓ_2, they may be parallel to each other, intersect each other at an angle not equal to 90°, or be perpendicular to each other. It is impossible for ℓ_1 to bisect ℓ_2 or vice-versa because they never end.

For 1–5, use the following diagram, in which lines ℓ_1 and ℓ_2 are parallel to each other and line ℓ_3 is a transversal.

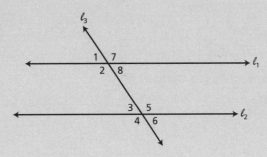

1.   Which one of the following angles does <u>not</u> have the same measure as $\angle 3$?

   (A) $\angle 1$ (C) $\angle 7$

   (B) $\angle 6$ (D) $\angle 8$

2.   Which one of the following pairs are called alternate exterior angles?

   (A) $\angle 2$ and $\angle 7$ (C) $\angle 4$ and $\angle 7$

   (B) $\angle 3$ and $\angle 8$ (D) $\angle 5$ and $\angle 6$

3.   If $m\angle 1 = 30°$, how many other angles have a measure of 30°?

   (A)   None   (C)   Two

   (B)   One   (D)   Three

Test Yourself! *(continued)*

4. What name is associated with the pair of angles $\angle 4$ and $\angle 2$?

(A) Vertical angles

(B) Alternate interior angles

(C) Corresponding angles

(D) Adjacent angles

5. If the measure of $\angle 2$ is 24° larger than the measure of $\angle 3$, what is the value of $m\angle 2$?

(A) 66° (C) 102°

(B) 78° (D) 156°

6. Given that $\overline{AC}$ bisects $\overline{BF}$ at point Z, consider the following three statements.

Statement I: $BZ = ZF$.

Statement II: $AZ = ZC$.

Statement III: All four angles at point Z have a measure of 90°.

Which of these statements must be true?

(A) Statements I, II, and III

(B) Only Statements I and II

(C) Only Statement II

(D) Only Statement I

Test Yourself! *(continued)*

7.   **Look at the following diagram:**

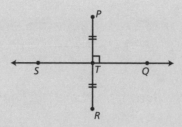

   Consider the following statements:

   Statement I: $\overline{PR}$ is perpendicular to $\overleftrightarrow{SQ}$

   Statement II: $PT = TR$

   Statement III: $ST = TQ$

   Which of these statements must be true?

   (A)   Only Statements I and II

   (B)   Only Statements II and III

   (C)   Only Statement I

   (D)   Only Statement III

8.   **Suppose you are given two parallel lines and a transversal for which $\angle 4$ and $\angle 6$ are adjacent angles. If $m\angle 6$ is nine times as large as $m\angle 4$, what is the value of $m\angle 4$?**

   (A)   18°   (C)   160°

   (B)   20°   (D)   162°

9.   Which one of the following represents an impossible situation?

(A)   Two lines that are perpendicular to each other

(B)   Two lines that bisect each other

(C)   Two line segments that are perpendicular to each other but do not bisect each other

(D)   A line that bisects a line segment but is not perpendicular to it

10.   Suppose you are given that point *M* is the midpoint of $\overline{RS}$. How many lines may intersect $\overline{RS}$ at point *M* and **not** be perpendicular to $\overline{RS}$?

(A)   Infinitely many   (C)   One

(B)   Two   (D)   None

QUIZ ONE

1. **Which one of the following is true concerning any two line segments?**

   A They must intersect in more than one point.

   B They might not intersect or be parallel to each other.

   C They must intersect or be parallel to each other.

   D They must be perpendicular or parallel to each other.

2. **Given that ∠1 and ∠2 are supplementary angles, which one of the following would imply that they are also congruent?**

   A ∠1 is an obtuse angle.

   B ∠1 and ∠2 are adjacent angles.

   C The measure of ∠2 is 90°.

   D ∠1 and ∠2 share a common vertex.

3. **For the following line segment, *FJ* = 40, *GH* = 15, and *HJ* = 12.**

   F ● —— G ● ———— H ● —— J ●

   What is the value of *FG*?

   A 10

   B 11

   C 12

   D 13

4. **Which one of the following is a correct description of a ray?**

   A It goes in two directions, with no end points.

   B It goes in one direction, with one end point.

   C It contains two end points.

   D It has no dimensions.

5. **Which one of the following pairs contains two acute angles?**

   A 50° and 150°

   B 100° and 200°

   C 68° and 95°

   D 10° and 87°

6. Consider the following diagram.

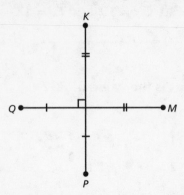

Which one of these statements is correct?

A $\overline{KP}$ is perpendicular to $\overline{QM}$, but $\overline{KP}$ does not bisect $\overline{QM}$.

B Each of $\overline{KP}$ and $\overline{QM}$ is the perpendicular bisector of the other segment.

C $\overline{KP}$ is the perpendicular bisector of $\overline{QM}$, but $\overline{QM}$ does not bisect $\overline{KP}$.

D $\overline{KP}$ bisects $\overline{QM}$, but $\overline{KP}$ is not perpendicular to $\overline{QM}$.

7. Consider the following diagram, in which line ℓ_1 is parallel to line ℓ_2.

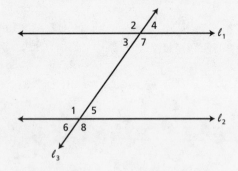

What name is assigned to a pair of angles such as $\angle 2$ and $\angle 8$?

A Alternate exterior angles

B Vertical angles

C Alternate interior angles

D Corresponding angles

8. For the following diagram, $\overline{RS}$ is 12 units longer than $\overline{ST}$. $RT = 84$ units.

What is the value of RS, in units?

A 54

B 48

C 42

D 36

9. Consider the following diagram.

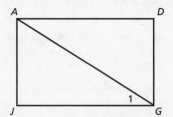

Which one of the following is an equivalent way to name $\angle 1$?

A $\angle GAJ$

B $\angle AGD$

C $\angle JGA$

D $\angle DAG$

10. For the following diagram, $m \angle NKL = 90°$, $m \angle QKM = 82°$, and $m \angle NKM = 57°$.

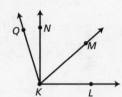

What is the measure of $\angle QKL$?

A 172°

B 147°

C 139°

D 115°

Perimeters of Polygons

In this lesson, we will talk about the perimeter of a polygon, which means that each figure will be made up of only line segments. Although not an absolute requirement, each angle of each figure is less than 180°. A real-life example that uses the concept of perimeter is the amount of fencing needed to enclose a backyard.

Your Goal: When you have completed this lesson, you should be able to find the perimeter of a polygon when given the length of each side. You should also be able to determine a specific length, given the value of the perimeter and other information.

LESSON 5

Perimeters of Polygons

Let's Review

SEE LESSON 1

The line segments that comprise each figure are called **sides**. Each side must be connected to exactly two other sides of the figure. An **angle** consists of two rays that share a common end point. In geometry, each angle is shortened to line segments that share a common end point. Look at the following diagram.

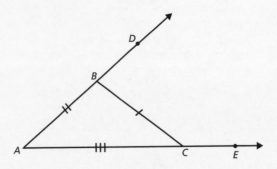

Technically, $\angle A$ is composed of rays $\overrightarrow{AD}$ and $\overrightarrow{AE}$. (Remember that we could also name these rays as $\overrightarrow{AB}$ and $\overrightarrow{AC}$.) Now, consider the same figure showing only the triangle.

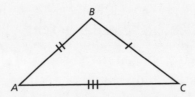

The different sets of tick marks show that the sides are not equal in length. For this polygon, (which you already know is a triangle), we can consider that $\angle A$ is composed of line segments (sides) $\overline{AB}$ and $\overline{AC}$. Basically, we are really just viewing part of $\angle A$. Thus, it will be acceptable to state that each angle of this triangle consists of two sides. Now let's discuss the meaning of **perimeter**.

The perimeter of any polygon is just the sum of the lengths of all its sides.
Any three-sided polygon is called a **triangle**.
A four-sided polygon is called a **quadrilateral**.
Another polygon we will use in this lesson is a **pentagon**, which has five sides.

1 Example: *What is the perimeter of a triangle with sides of lengths 14 feet, 17 feet, and 23 feet?*

Solution: The perimeter is 14 + 17 + 23 = 54 feet.

2 Example: *Each side of a pentagon is 22 units. What is the perimeter?*

Solution: The perimeter is (5)(22) = 110 units.

3 Example: *If two sides of a triangle are 5 inches and 7 inches and the perimeter is 20 inches, what is the length of the third side?*

Solution: Let x represent the length of the third side. Then $5 + 7 + x = 20$. This equation can be simplified to $12 + x = 20$, so $x = 8$ inches.

4 Example: *One side of a triangle is 10 units. The second side is 3 units longer than the third side.*
If the perimeter is 55 units, what is the length of the longest side?

Solution: Let x represent the length of the third side, and let $x + 3$ represent the length of the second side.
$$10 + (x + 3) + x = 55.$$
Simplify to $2x + 13 = 55$.
$$2x = 55 - 13.$$
$$2x = 42, \text{ so } x = 21.$$
Don't stop here! The second side, whose length is 24 units, is the longest side.

MathFlash!

In example 4, you could let x represent the length of the second side, so that x – 3 would represent the length of the third side. Then, the value of x would be the required answer.

5 **Example:** *In a quadrilateral, the lengths of three of the sides are 7 cm, 29 cm, and 24 cm.*
If the perimeter is 71 cm, what is the length of the fourth side?

Solution: Let x represent the unknown length. Then $x + 7 + 29 + 24 = 71$. This equation simplifies to $x + 60 = 71$, so $x = 11$ cm.

6 **Example:** *In a pentagon, the lengths of three of the sides are 80 units, 52 units, and 16 units. The fourth side is five times as large as the fifth side. The perimeter is 238 units.*

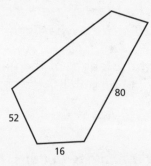

What is the length of the shortest side of this figure?

Solution: Let x represent the length of the fifth side, and let $5x$ represent the length of the fourth side.
Then $80 + 52 + 16 + 5x + x = 238$.
Simplify to get $148 + 6x = 238$. Then $6x = 90$, so $x = 15$.
This means that 15 units is the length of the fifth side, and $(5)(15) = 75$ is the length of the fourth side.
The question asked for the length of the shortest side, so our answer is 15.

MathFlash!

The sum of the lengths of any two sides of a triangle must be more than the length of the third side.
For a quadrilateral, the sum of the lengths of any three sides must be more than the length of the fourth side.
Finally, for a pentagon, the sum of the lengths of any four sides must exceed the length of the fifth side.

7 Example: *Which one of the following groups of numbers cannot represent the lengths of the three sides of a triangle?*

(A) 4, 5, 7   *(C) 6, 9, 20*

(B) 9, 15, 15   *(D) 7, 7, 7*

Solution: The correct answer is (C) since 6 + 9 < 20. The quickest way to determine if the three numbers do not form a triangle is to simply add the two smallest numbers. This sum must be more than the size of the largest side.
It is possible for any two sides or even all three sides to be the same size.

8 Example: *Four sides of a pentagon are 18, 13, 80, and 30. Which one of the following numbers could represent the length of the fifth side?*

(A) 150   *(C) 19*

(B) 25   *(D) 5*

Solution: The correct answer is (B). The sum of the four given sides is 141, so answer choice (A) must be wrong. Answer choice (C) is wrong because 13 + 18 + 19 + 30 = 80. (Remember that the sum of any four sides must be greater than the fifth side.) Answer choice (D) is wrong because 5 + 13 + 18 + 30 < 80.
Note that with answer choice (B), we have 13 + 18 + 25 + 30 > 80.

9 Example: *Given triangle GHJ, ∠H is composed of which two line segments?*

Solution: Just select the two sides that have an end point of *H*. The answers are $\overline{GH}$ and $\overline{HJ}$. The actual figure could look like:

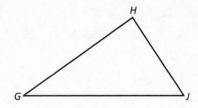

10   **Example:**   *Given quadrilateral PQRS, which angle is composed of $\overline{RS}$ and $\overline{SP}$?*

Solution: The common end point of these two sides is *S*. Thus, the answer is $\angle S$. The actual figure could appear as:

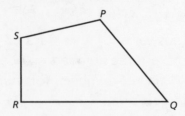

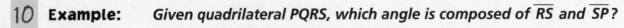

MathFlash!

These polygons are named by using end points of line segments. These end points are called "vertices." (The singular is "vertex.") Any polygon with at least four sides is named by starting at any vertex and moving around the figure in either direction. Usually, the lettering is alphabetical.

Thus, MNPQR, PQRMN, and QPNMR are three acceptable ways to name the following polygon.

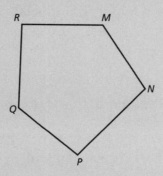

However, both *MNQRP* and *NRPQM* would <u>not</u> be acceptable in naming the above polygon because the letters are not in order.

1.   **Which one of the following would not be classified as a polygon?**

(A)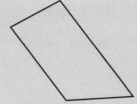

(C)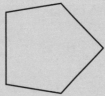

(B)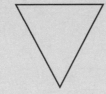

(D)

2.   **Look at the following quadrilateral.**

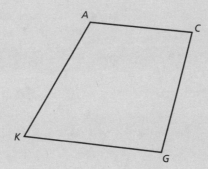

Which one of the following is <u>not</u> an acceptable way of naming it?

(A) *KGCA*   (C) *ACGK*

(B) *CKGA*   (D) *GKAC*

3. Each side of a quadrilateral is 7 inches. What is its perimeter?

(A)   14 inches

(C)   28 inches

(B)   21 inches

(D)   49 inches

4. A pentagon has a perimeter of 180 units. If each side has the same length, what is the length of one side?

(A)   60 units

(C)   36 units

(B)   45 units

(D)   30 units

5. One side of the triangle is 50 cm. A second side is three times as large as a third side.

50 cm

If the perimeter is 122 cm, what is the length of the largest side?

(A) 72 cm

(C)   50 cm

(B) 54 cm

(D)   18 cm

6. Two sides of a triangle are 17 units and 11 units. Which one of the following could be the length, in units, of the third side?

(A) 36

(C)   28

(B) 32

(D)   10

 **Test Yourself!** (continued)

7. Which one of the following <u>cannot</u> represent the lengths of the sides of a quadrilateral?

    (A) 3, 5, 7, 16   (C) 80, 70, 60, 10

    (B) 4, 8, 12, 20   (D) 32, 31, 2, 1

8. For any polygon, each vertex must belong to how many of the sides?

    (A) One   (C) Three

    (B) Two   (D) Four

9. In the quadrilateral, two of the sides have lengths of 5 inches and 12 inches. The third side is 9 inches longer than the fourth side.

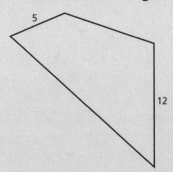

    If the perimeter is 42 inches, what is the length of the longest side?

    (A) 12 inches   (C) 17 inches

    (B) 14 inches   (D) 21 inches

10. In a pentagon, three of the sides are 8 units, 10 units, and 27 units. The remaining two sides are both equal to each other and are the shortest sides of this pentagon. Which of the following could be the length of each shortest side?

    (A) 2 units   (C) 4 units

    (B) 3 units   (D) 5 units

Triangles—Part I

In this lesson, we will explore the properties of different categories of triangles. Judging by how often we see them in everyday life, as well as on geometry exams, triangles are very popular figures. Here are just a few real-life situations involving triangles: (a) a person casting a shadow, (b) a three-sided fence in a backyard, (c) a ladder leaning against a building, and (d) a "yield" traffic sign.

Your Goal: When you have completed this lesson, you should be able to identify various types of triangles, as well as understand the relationships that exist among the angles and sides of triangles.

LESSON 6

Triangles—Part 1

We can classify triangles by the sizes of their angles. It is important to know that **the sum of the measures of all three angles in any triangle is always 180°.**

An **acute** triangle is one in which the measure of each of its angles is less than 90°. A **right** triangle is one which the measure of one of its angles is exactly 90°. Since the sum of all three angles is 180°, the measure of each of the other two angles must be less than 90°.

An **obtuse** triangle is one in which the measure of one of its angles is greater than 90°. By necessity, the measure of each of the other two angles must be less than 90°. Here are diagrams to illustrate these three types of triangles.

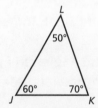

Figure 6.1
Acute Triangle

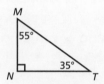

Figure 6.2
Right Triangle

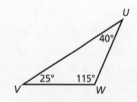

Figure 6.3
Obtuse Triangle

There are additional categories of triangles that are dependent on the relationship among the sides (and therefore among the angles). Let's first introduce the word **congruent**. If the measure of ∠A is equal to the measure of ∠B, we can state that ∠A is congruent to ∠B. If PQ = RS, we can state that $\overline{PQ}$ is congruent to $\overline{RS}$.

A **scalene** triangle is one in which all three sides have different lengths. It is also true that the measures of all three angles will be different. In Figures 6.1, 6.2, and 6.3 shown above, each triangle is scalene.

An **isosceles** triangle is one in which at least two sides are **congruent**. If a problem states that a triangle is isosceles, you can only assume that two sides are congruent. Here are some examples of isosceles triangles.

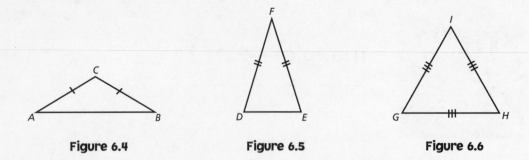

Figure 6.4   Figure 6.5   Figure 6.6

Remember, whenever you see the same number of tick marks for sides (or angles), it means that the marked sides (or angles) are congruent. In Figures 6.4 and 6.5, exactly two sides are congruent. Note that in Figure 6.6, all three sides are congruent. When this occurs, the triangle is called **equilateral**. You can think of an equilateral triangle as <u>a special case</u> of an isosceles triangle.

At this point, use your protractor to measure each angle of Figures 6.4, 6.5, and 6.6. Hopefully, you are not surprised to discover the following:

In Figure 6.4, $m\angle A = m\angle B = 30°$ and $m\angle C = 120°$.

In Figure 6.5, $m\angle D = m\angle E = 75°$ and $m\angle F = 30°$.

In Figure 6.6, each angle has a measure of 60°.

In any isosceles triangle, such as Figures 6.4 and 6.5, the two congruent angles are called **base** angles.

In Figure 6.6, any two of the three angles can be considered base angles. In effect, we are stating that in any isosceles triangle, at least two angles (base angles) must be congruent.

In Figures 6.4 and 6.5, the angle whose measure is different from the other two angles is called the **vertex** angle.

In Figure 6.6, any of the three angles can be considered as a vertex angle. Incidentally, an equilateral triangle is also called an **equiangular** triangle. You can probably guess why!

Let's redraw Figures 6.4, 6.5, and 6.6 by using tick marks to show congruent sides and congruent angles. We'll call these diagrams Figures 6.4a, 6.5a, and 6.6a.

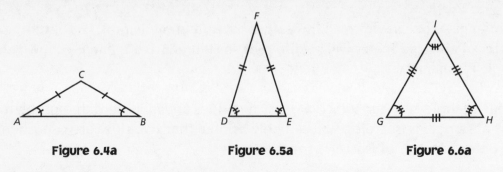

Figure 6.4a   Figure 6.5a   Figure 6.6a

At this point, we are going to return to Figures 6.1 through 6.5, each of which has been purposely drawn to scale. For the triangles in Figures 6.1, 6.2, and 6.3, use your ruler to identify the smallest, next largest, and largest sides. Look at your results.

For Figure 6.1, you should have discovered that $\overline{JK}$ is the smallest segment and $\overline{JL}$ is the largest segment. This implies that $JK < KL < JL$.
For, Figure 6.2, $\overline{MN}$ is the smallest segment, and $\overline{MT}$ is the largest segment. This means that $MN < NT < MT$.
For Figure 6.3, you probably found that $\overline{UW}$ is the smallest segment and $\overline{UV}$ is the largest segment. Thus, it is true that $UW < VW < UV$.

You may be wondering where all this information is leading. Don't worry; we are not traveling down a dark and deserted road leading to a dead end! Looking again at Figure 6.1, notice that the smallest side ($\overline{JK}$) is opposite the smallest angle ($\angle L$). Also, the largest side ($\overline{JL}$) is opposite the largest angle ($\angle K$). Notice that $\overline{KL}$ is neither the largest nor the smallest side, and thus the angle opposite $\overline{KL}$ ($\angle J$) is neither the largest nor the smallest angle. Does this type of reasoning hold for Figures 6.2 and 6.3? Absolutely!

Look at Figure 6.4 . We already know that $\angle C$ is the largest angle. The largest side is $\overline{AB}$. Notice that $\overline{AB}$ lies opposite $\angle C$. Since $\angle A$ and $\angle B$ are congruent, it should come as no surprise that the sides opposite these angles ($\overline{BC}$ and $\overline{AC}$) are also congruent.

Can you now see that in Figure 6.5, since $\angle F$ is the smallest angle, $\overline{DE}$ must be the smallest side? $\angle D$ and $\angle E$ are congruent, so we expect that $\overline{DF}$ and $\overline{EF}$ are also congruent. The rule that we are establishing is as follows: **Within a given triangle, the largest angle always lies opposite the largest side. Also, the smallest angle always lies opposite the smallest side.**

MathFlash!

Without a diagram, the easiest way to identify a side opposite a given angle in a triangle is to simply use the "other" two vertices to represent the side. For example, in triangle ACY, the side opposite $\angle C$ is $\overline{AY}$. Of course, this can be reversed. If you are given triangle BHM, the angle opposite $\overline{HM}$ must be $\angle B$.

CAUTION

Do <u>not</u> use the rule we have just established if you are comparing sides and angles from different triangles! It is certainly easy to construct two triangles in which the side opposite a large angle in one triangle is smaller than a side opposite a smaller angle in the other triangle. Figures 6.7 and 6.8 illustrate this concept.

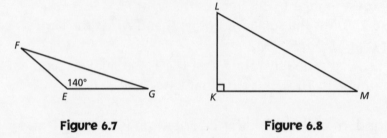

Figure 6.7   Figure 6.8

In triangle *EFG*, m∠*E* = 140°. In triangle *KLM*, m∠*K* = 90°. However, $\overline{FG}$ is smaller than $\overline{LM}$.

1 **Example:** *Triangle XYZ is an isosceles right triangle, with m∠Z = 90°. What is the measure of ∠X?*

Solution: Here is an appropriate diagram, although you don't really need it.

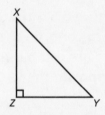

Since this triangle is isosceles, at least two angles must be congruent. But you know that <u>only</u> two angles are congruent: ∠*X* and ∠*Y*. (It is important to remember that if all three angles are congruent in a triangle,then the measure of each one must be 60°.) Let *x* represent the measure of each of ∠*X* and ∠*Y*. Then 90 + *x* + *x* = 180. 2*x* = 180 – 90. This simplifies to 2*x* = 90, so *x* = 45°. (Incidentally, this is also the measure of ∠*Y*.)

2 **Example:** *In a particular triangle, the measure of the first angle is 36°. The measure of the second angle is three times the measure of the third angle.*
What is the measure of the largest angle of this triangle?

Solution: Let *x* represent the measure of the third angle, and let 3*x* represent the measure of the second angle.
Then $36 + x + 3x = 180$.
This simplifies to $4x + 36 = 180$.
Then $4x = 144$, so $x = 36°$.
The measure of the third angle is $(3)(36) = 108°$,
so this is the largest angle.

3 **Example:** *What two labels can be assigned to the type of triangle in Example 2?*

Solution: Since the angles are 36°, 36°, and 108°, we can label the triangle as obtuse and as isosceles.

4 **Example:** *What is an illustration of the angle values assigned to an acute triangle that is also scalene, where the measure of the largest angle is 65°?*

Solution: There will be several answers for this question.
The sum of the other two angles must be $180 - 65 = 115°$.
At this point, you could use any two numbers under 90.
But they have to add up to 115.
One possibility for the measures of the missing angles is 56° and 59°.
The measures of the three angles would then be 56°, 59°, and 65°, and the triangle could appear as follows:

5   **Example:**   *In right triangle RST, ∠T is a right angle. If the measure of ∠R is 24° greater than the measure of ∠S, what is the measure of the smallest angle?*

Solution: Let x represent the measure of ∠S, and let $x + 24$ represent the measure of ∠R.

Then $x + x + 24 + 90 = 180$.

This simplifies to $2x + 114 = 180$.

Then $2x = 66$, so $x = 33°$, which is the measure of ∠S.

(You would not have to find the measure of ∠R, since it cannot be the smallest angle. However, here it is, $m∠R = 57°$. Note that the sum of the measures of all three angles is 180°.)

6   **Example:**   *What are the three angle values of a triangle that is isosceles, obtuse, and whose smallest angle measures 14°?*

Solution: At first, it may seem as if there are several answers. But notice that we have an obtuse triangle, so the measure of one of the angles must be greater than 90°. Since this triangle is also isosceles, two angles must be congruent. Therefore, the measures of the two smallest angles of this triangle must be 14°. The measure of the third angle is $180 - 14 - 14 = 152°$. The triangle could appear as follows:

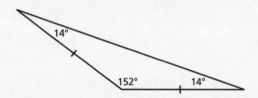

7   **Example:**   *Which <u>two</u> of the following selections are <u>impossible</u> to construct?*

   (A) An equilateral triangle with the measure of one angle as 70°

   (B) An acute triangle with an angle whose measure is less than 10°

   (C) An obtuse triangle with an angle whose measure is less than 10°

   (D) An isosceles triangle in which the measure of one of the congruent angles is 95°

   (E) A scalene triangle that also has a right angle

Solution: The correct answers are (A) and (D).

Answer choice (A) is impossible because in every equilateral triangle, no matter its size, the measure of each angle is 60°.

Answer choice (B) is possible. An example would be a triangle with angle measures of 8°, 85°, and 87°.

Answer choice (C) is possible. An example would be a triangle with angle measures of 6°, 54°, and 120°.

Answer choice (D) is impossible because if a triangle had two 95° angles, their sum would already exceed 180°.

Answer choice (E) is possible. An example would be a triangle with angle measures of 35°, 55°, and 90°.

8 **Example:**   *Look at the following diagram:*

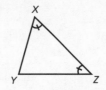

Which one of the following correctly describes this triangle?

(A) An obtuse triangle   *(C) A right triangle*

(B) An isosceles triangle   *(D) An equiangular triangle*

Solution: The correct answer is (B). By the markings shown, $\angle X$ and $\angle Z$ are congruent angles. This implies that $XY = YZ$, so the triangle must be isosceles. For this reason, answer choices (A) and (C) are incorrect because we are not given the size of $\angle Y$. Hopefully, you realized that answer choice (D) is totally wrong. An equiangular triangle must have three 60° angles.

9 Example: *In triangle ABC, AB = 10 , BC = 12, and AC > AB.*
Which one of the following <u>must be</u> true?

(A) *AC > 12*   (C) *∠C is the smallest angle*

(B) *$\overline{AC}$ is the largest side*   (D) *∠A is the largest angle*

Solution: The correct answer is (C). Since $AC > AB$, we know that $AC > 10$. This
implies that the shortest side is $\overline{AB}$. Then the smallest angle must lie
opposite $\overline{AB}$, which is $\angle C$.

Answer choices (A) and (B) are <u>not</u> absolutely guaranteed. We can
determine that the length of $\overline{AC}$ can be any value between 10 and
22 (not inclusive); thus $\overline{BC}$ could actually be the largest side.
Answer choice (D) would be correct only if we can guarantee that
$\overline{BC}$ is the largest side.

*There is a commonly accepted symbol for the word "triangle"
when referring to a specifically named triangle. Instead of writing
"triangle ABC," we can write "△ABC." The symbol △ is used, even if
the figure is a right triangle or an obtuse triangle. This symbol will
be used in the following Drill Exercises.*

Test Yourself!

1.   **Which one of the following could represent two of the three angles
of an acute triangle?**

(A) **28° and 66°**   (C) **60° and 10°**

(B) **4° and 100°**   (D) **31° and 59°**

2.   In △LMN, LM = 5, LN = 7, and MN = 9. Which angle is the smallest?

      Answer: _____

3.   In △STU, m∠S = 40° and m∠T = 35°. Which one of the following <u>must</u> be true?

      (A) △STU has a right angle.   (C) △STU is isosceles.

      (B) ∠S is the largest angle.   (D) $\overline{ST}$ is the largest side.

4.   Which one of the following is the <u>minimum</u> requirement that would <u>guarantee</u> that a triangle is equilateral?

      (A) One of its angles is 60°.

      (B) Two of its angles are each 60°.

      (C) All three angles are each 60°.

      (D) None of the above

5.   You are given two triangles, △DEF and △HIJ, for which m∠D = 50° and m∠H = 70°. When you measure $\overline{EF}$ and $\overline{IJ}$, you discover that EF > IJ. Which of the following situations must be present?

      (A) △DEF is larger than △HIJ.

      (B) △DEF is smaller that △HIJ.

      (C) Both triangles are acute.

      (D) One of these is a right triangle.

6.   In an isosceles triangle, one of the congruent angles has a measure of 41°. What is the measure of the largest angle?

(A)   49°   (C)   98°

(B)   82°   (D)   139°

7.   In $\triangle PQR$, $m\angle P = 54°$. The measure of $\angle Q$ is six times as large as the measure of $\angle R$? What is the value of $m\angle Q$?

(A)   18°   (C)   108°

(B)   21°   (D)   126°

8.   Which of the following is the best description of a triangle for which two angle measures are 45° and 90°?

(A)   Isosceles triangle that is acute

(B)   Right triangle that is isosceles

(C)   Scalene triangle that is obtuse

(D)   Right triangle that is scalene

9.   Which one of the following <u>must</u> be an acute triangle?

(A)   An equilateral triangle

(B)   A triangle that is scalene

(C)   A triangle in which two of the angles measure 35° and 45°

(D)   An isosceles triangle

10.   In $\triangle JKL$, $m\angle J = 58°$. The measure of $\angle K$ is 12° greater than the measure of $\angle L$? Which one of the following inequalities is correct?

(A) $m\angle J < m\angle K < m\angle L$ (C) $m\angle J < m\angle L < m\angle K$

(B) $m\angle L < m\angle K < m\angle J$ (D) $m\angle L < m\angle J < m\angle K$

Triangles—Part 2

In this lesson, we will explore additional properties of triangles. Some of these properties deal with different line segments and angles that are not part of the actual triangle. You may want to review Lesson 4, since we will discuss properties that involve both parallel lines and triangles.

Your Goal: When you have completed this lesson, you should be able to understand the relationships among the angles in figures containing both parallel line segments and a triangle. You will also be able to classify other line segments that are associated with triangles.

LESSON 7

Triangles—Part 2

Consider Figure 7.1, involving △ABC, in which $\overline{BC}$ is part of $\overrightarrow{BD}$.

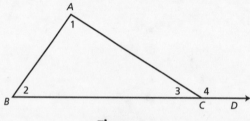

Figure 7.1

∠4 is called an **exterior** angle of △ABC. ∠3 is called an **adjacent interior** angle to ∠4, because together they form a straight line. Again, relative to ∠4, each of ∠1 and ∠2 is called a **remote interior** angle. You already know that since ∠3 and ∠4 are adjacent angles that form a line, the sum of their measures is 180°. But we also know that the sum of the measures of ∠1, ∠2, and ∠3 is also 180°.

If we substitute, we can see that $m\angle 1 + m\angle 2 = m\angle 4$. This implies that **the measure of an exterior angle of a triangle is equal to the sum of the measures of its two remote interior angles.**

For any triangle, it is possible to construct a total of six different exterior angles, each of which has a measure equal to the sum of the measures of its two associated remote interior angles. Figures 7.2, 7.3, 7.4, 7.5, and 7.6 illustrate these possibilities for △ABC. In each case, the measure of the exterior angle is equal to the sum of the measures of its two associated interior angles.

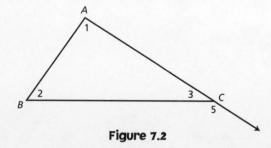

Figure 7.2

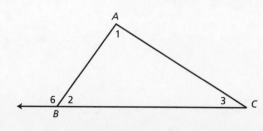

Figure 7.3

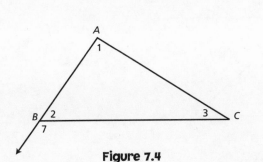

Figure 7.4

Figure 7.5

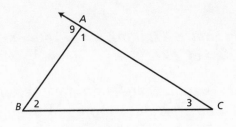

Figure 7.6

1 **Example:** *Using Figure 7.1, if m∠1 = 100° and m∠4 =163°, what is the measure of ∠2?*

Solution: Let *x* represent the measure of ∠2. Then $x + 100 = 163$, so $x = 63°$. Note that $m∠3 = 180 - m∠4 = 17°$. Sure enough, we can verify that $m∠1 + m∠2 + m∠3 = 180°$.

2 **Example:** *Using Figure 7.4, if m∠1 = 95° and m∠3 = 16°, what is the measure of ∠7?*

Solution: There is no need to use any fancy steps. $m∠7 = 95 + 16 = 111°$.

Now consider Figure 7.7, involving parallel lines ℓ_1 and ℓ_2. Each of $\overline{EF}$ and $\overline{EG}$ is a transversal. We can identify two pairs of congruent angles. Let's show each of these pairs, using the notation ≅. Using the transversal $\overline{EF}$, we recognize that ∠1 and ∠5 are alternate interior angles of parallel lines and are therefore congruent. We can state this as $m∠1 = m∠5$ or as $∠1 ≅ ∠5$. Also, using $\overline{EG}$ as a transversal, we can write $m∠3 = m∠6$ or $∠3 ≅ ∠6$.

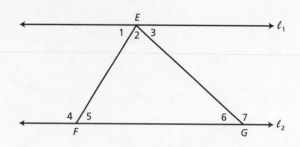

Figure 7.7

3   **Example:**   *Using Figure 7.7, if m∠3 = 37° and m∠5 = 68°, what is the measure of ∠2?*

Solution: Since ∠1 and ∠5 are alternate interior angles of parallel lines, $m\angle 1 = 68°$. Now let x represent the measure of ∠2.
Since ∠1, ∠2, and ∠3 form a straight line,
$m\angle 1 + m\angle 2 + m\angle 3 = 180°$.
By substitution, $68 + x + 37 = 180°$.
Then $x = 75°$.

4   **Example:**   *Using Figure 7.7, if m∠7 = 125° and m∠2 = 73°, what is the measure of ∠1?*

Solution: Immediately, we know that $m\angle 6 = 180 - 125 = 55°$. Since ∠3 and ∠6 are alternate interior angles of parallel lines, $m\angle 3 = 55°$.
Finally, $m\angle 1 + m\angle 2 + m\angle 3 = 180°$.
Substitute the known values with x representing the measure of ∠1.
Then $x + 73 + 55 = 180$, which leads to $x + 128 = 180$.
Finally, $x = 52°$.

MathFlash!

Remember that a geometric drawing may not be drawn to scale. It is more important to check your answer algebraically. However, when you see drawings in this book, the angle measures given will be very close to the measures obtained by using a protractor. For example, if an angle measure is stated as less than 90°, if you used a protractor, the angle would actually be acute.

Let's introduce a few new names for line segments related to a triangle.
An **altitude** of a triangle is the segment that is drawn from one vertex to the line
containing the opposite side such that it is perpendicular to the side to which it is
drawn. There are three altitudes in every triangle, one stemming from each vertex.
Look at the following figures of triangles with their altitudes.

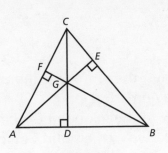

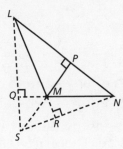

| Figure 7.8 | Figure 7.9 | Figure 7.10 |

In Figure 7.8, $\triangle ABC$ is acute. The three altitudes are $\overline{AE}$, $\overline{CD}$, and $\overline{BF}$. Each of the three
altitudes lies within the interior of $\triangle ABC$, and they intersect at a common point G.
In Figure 7.9, $\triangle HIJ$ is a right triangle. You can see that $\overline{IK}$ is an altitude drawn from
the vertex labeled *I* to $\overline{HJ}$. The other two altitudes are already in the drawing! Since
$\overline{HI}$ is perpendicular to $\overline{IJ}$, $\overline{HI}$ is an altitude. Also, $\overline{IK}$ is perpendicular to $\overline{HJ}$, so $\overline{IK}$ is an
altitude. Notice that the three altitudes intersect at a common point, namely, *I*.

Before we inspect Figure 7.10, consider the following obtuse triangle *XYZ*, in which
$\angle YXZ$ is an obtuse angle.

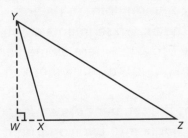

In order to draw the altitude from point *Y*, we must extend $\overline{XZ}$ to a point *W* such that
$\overline{YW}$ is perpendicular to $\overline{XW}$. Each of $\overline{YW}$ and $\overline{XW}$ has been shown with dotted lines to
emphasize that they are not part of $\triangle XYZ$.

In Figure 7.10, $\triangle LMN$ is obtuse, with its obtuse angle at *M*. $\overline{MP}$ is an altitude, since $\overline{MP}$
is perpendicular to $\overline{LN}$. In order to draw the altitude from point *L*, we realize that any
line segment drawn from *L* to the segment $\overline{MN}$ would never create a right angle. The
method in which an altitude is drawn from point L is to extend $\overline{MN}$ to a point Q such
that $\angle LQM = 90°$. $\overline{LQ}$ is considered the altitude from point *L*. The same scene occurs
at point *N*. If you were to connect *N* to any point on $\overline{LM}$, you would not be able to
create a right angle. Thus, $\overline{LM}$ is extended to a point R such that $\angle NRM = 90°$. $\overline{NR}$ is the
altitude from point *N*. In this type of triangle, the three altitudes, when extended far
enough, will intersect at a common point *S*.

For acute and right triangles, the three altitudes intersect in the interior of a triangle. For an obtuse triangle, the extensions of the altitudes do intersect at a common point that lies outside the triangle.

The length of any altitude of a triangle is called its **height**. Thus, the altitude is the actual line segment, whereas the height is associated with its numerical length. A **median** of a triangle is the segment drawn from one vertex to the midpoint of the opposite side. There are three medians in every triangle, as shown in the figures below.

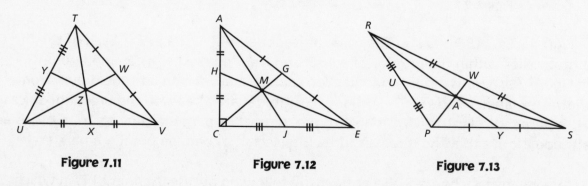

| Figure 7.11 | Figure 7.12 | Figure 7.13 |

Figures 7.11, 7.12, and 7.13 show acute, right, and obtuse triangles, respectively. Note that the medians intersect at a common point in the interior of the given triangle. The indicated tick marks show that the drawn segments are medians. For example, in Figure 7.11, $\overline{TX}$ is a median from point T to $\overline{UV}$, where all medians intersect at point Z. In Figure 7.12, $\overline{CG}$ is a median from point C to $\overline{AE}$, where all medians intersect at point M.

An **angle bisector** of a triangle is the segment drawn from one vertex to the opposite side so that it divides the vertex angle into two congruent angles. There are three angle bisectors in every triangle, as shown in the figures below.

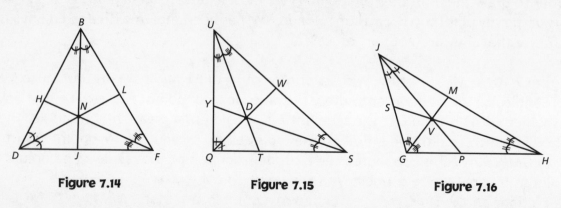

| Figure 7.14 | Figure 7.15 | Figure 7.16 |

Figures 7.14, 7.15, and 7.16 show acute, right, and obtuse triangles, respectively. Note that the angle bisectors intersect at a common point in the interior of the given triangle. The indicated tick marks on the original angles of the triangle indicate that the drawn segments are angle bisectors. For example, in Figure 7.15, $\overline{UT}$ is an angle bisector from $\angle U$ to $\overline{QS}$. In Figure 7.16, $\overline{JP}$ is an angle bisector from $\angle J$ to $\overline{GH}$.

A **midsegment** of a triangle is the segment drawn between the midpoints of any two sides. There are three midsegments in every triangle, as shown in the figures below. Figures 7.17, 7.18, and 7.19 show acute, right, and obtuse triangles, respectively.

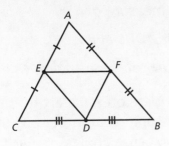

Figure 7.17

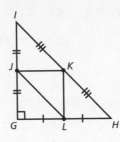

Figure 7.18

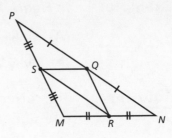

Figure 7.19

A midsegment of a triangle is parallel to the side that is not connected to it. The indicated tick marks of the three sides of the given triangle indicate that the drawn segments are midsegments. For example, in Figure 7.17, $\overline{EF}$ is parallel to $\overline{BC}$. In Figure 7.19, $\overline{QR}$ is parallel to $\overline{PM}$.

Before ending this lesson, take your ruler and measure some of the figures:

Measure $\overline{ED}$ and $\overline{AB}$ in Figure 7.17.
Then go to Figure 7.18 and measure both $\overline{JK}$ and $\overline{GH}$.
Finally, in Figure 7.19, measure both $\overline{QR}$ and $\overline{PM}$.
Did you notice any pattern for each pair of segments? If it seemed that the midsegment was about half the size of the opposite side of the triangle, that is exactly the correct conclusion. Thus, in any triangle, a midsegment is parallel to and one half the length of the third side. (The "third side" is simply the side of the triangle that does not intersect the given midsegment.)

Let's use the figures in this lesson for the remaining examples.

5 **Example:**   *In Figure 7.10, what is the name of the altitude drawn from point L?*

Solution: The altitude from point L is $\overline{LQ}$.

6 **Example:** *In Figure 7.8, if the height of the altitude from point B is 16, which of the following is correct?*

(A) *BG = 16*   (C) $\overline{FG}$ *= 16*

(B) *BF = 16*   (D) $\overline{FB}$ *= 16*

Solution: The correct answer is (B). The referenced altitude is $\overline{FB}$ (or $\overline{BF}$). However, the height must be a number, which is indicated by *BF*.

7 **Example:** *In Figure 7.15, which angle must have a measure of 45°?*

(A) *Only ∠QUS*   (C) *Both ∠QUS and ∠QSU*

(B) *Only ∠QSU*   (D) *Neither of these angles*

Solution: The correct answer is (D). Hopefully, you were on full alert for this question! Remember that a right triangle only guarantees one 90° angle. In order to be assured that the two acute angles are each 45°, the triangle must also be isosceles.

8 **Example:** *In Figure 7.19, if SQ = 15, which of the following segments must also have a length of 15?*

(A) $\overline{RN}$ (C) $\overline{PQ}$

(B) $\overline{SM}$ (D) $\overline{RQ}$

Solution: The correct answer is (A). A midsegment is half the length of the opposite side. Thus, $SQ = \left(\dfrac{1}{2}\right)(MN)$. Then *MN* must be 30, and each of $\overline{MR}$ and $\overline{RN}$ must have a length of 15.

9 **Example:**   *In Figure 7.18, which <u>two</u> of the following angles must have the same measure as ∠H?*

(A) ∠I

(D) ∠IKJ

(B) ∠JLG

(E) ∠HKL

(C) ∠JLK

Solution: The correct answers are (B) and (D). Since $\overline{JL}$ is a midsegment, it is parallel to $\overline{IH}$. Using $\overline{GH}$ as a transversal, ∠JLG and ∠H are corresponding angles, so they are congruent.
Similarly, $\overline{JK}$ is a midsegment and is parallel to $\overline{GH}$. Using $\overline{IH}$ as a transversal, ∠H and ∠IKJ are corresponding angles, so they are congruent.
None of the angles mentioned in answer choices (A), (C), and (E) are necessarily congruent to ∠H.

Test Yourself!

1.   **Look at the following figure:**

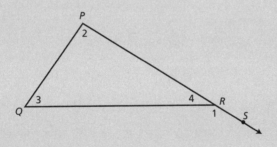

Relative to ∠1, ∠3 is considered a _____ _____ angle.

2.   **Using the figure in exercise 1, if *m*∠3 = 56° and the measure of ∠2 is 25° greater than that of ∠3, what is the measure of ∠1?**

Answer:_____

3.   **Look at the following figure:**

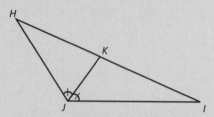

Which one of the following statements must be true?

(A)   *HK = IK*

(B)   *m∠HJK = m∠KJI*

(C)   If the altitude from point *H* lies outside the triangle, then *m∠HJI* <90°.

(D) $\overline{JK}$ is perpendicular to $\overline{HI}$.

4.   **For which type of triangle must all three altitudes intersect at a point outside the triangle?**

(A)   Scalene   (C)   Obtuse

(B)   Right   (D)   Equilateral

5.   **Look at the following figure in which lines ℓ_1 and ℓ_2 are parallel:**

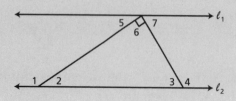

If *m∠4* = 116°, what is the measure of ∠5?   *Answer:*_____

Test Yourself! *(continued)*

6.   Using the figure in question 5, how many different angles are complementary to ∠7?

   (A)   None   (C)   Two

   (B)   One   (D)   Three

7.   For which type of triangle do all three medians intersect inside the triangle?

   (A)   Only right   (C)   Only right and acute

   (B)   Only acute   (D)   Every type

8.   Look at the following figure. Which one of the following angles is congruent to ∠*IKG*?

   (A)   ∠*IGK*

   (B)   ∠*E*

   (C)   ∠*A*

   (D)   ∠*KIG*

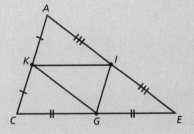

   ∠*IKG* must be congruent to which other angle?   *Answer:*_____

9.   Using the figure in question 8, if *AE* = 12, how many different segments must have a length of 6?

   (A)   Four   (C)   Two

   (B)   Three   (D)   One

10.   Using the figure in question 8, if *m*∠*IGE* = 65°, name two other angles whose measure must be 65°? (There are actually three answers.)

   *Answers:*_____

Triangles—Part 3

In this lesson, we will explore additional properties of the different line segments that are not part of the actual triangle. Be sure that you are thoroughly familiar with the key ideas covered in Lesson 7, especially those involving "altitude," "median," "angle bisector," and "midsegment." We will use these terms in connection with the concept of area.

Your Goal: When you have completed this lesson, you should be able to understand the relationship between each of the four key concepts mentioned above and the sides of a triangle. You will also be able to identify the connection between area and these various line segments.

LESSON 8

Triangles—Part 3

Consider Figure 8.1, involving △DEF, with its three altitudes.

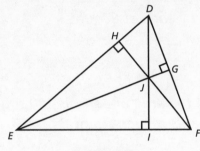

Figure 8.1

The **area** of any geometric figure is the number of square units of the region enclosed by the figure. In particular, the area of a triangle is given by the formula $A = \frac{1}{2}bh$, where A = area, b = any of the three sides, and h = the length of the corresponding altitude drawn to the selected side. The letter b is used because the selected side can be called the base. The letter h is used to represent the height.

For Figure 8.1, the area of the triangle could be found by using any one of the following: $\left(\frac{1}{2}\right)(DE)(HF)$, $\left(\frac{1}{2}\right)(EF)(DI)$, or $\left(\frac{1}{2}\right)(DF)(GE)$. These formulas will be used with any type of triangle (acute, obtuse, isosceles, etc.).

1 **Example:** *Given that EF = 8 and DI = 5, what is the area of △DEF?*

Solution: By substitution, $A = \left(\frac{1}{2}\right)(8)(5) = 20$.

When you multiply the numbers $\frac{1}{2}$, 8, and 5, you can do this in any of three ways:

(a)   Multiply $\frac{1}{2}$ by 8 to get 4, and then multiply 4 by 5.

(b)   Multiply 8 by 5 to get 40, and then multiply 40 by $\frac{1}{2}$.

(c)   Multiply $\frac{1}{2}$ by 5 to get 2.5, and then multiply 2.5 by 8.

Each of (a), (b), and (c) will yield the correct answer of 20. But do <u>not</u> multiply $\frac{1}{2}$ by each of 8 and 5.

2   Example: *Given that the area of △DEF is 100 square inches and FH = 30 inches, what is the length of $\overline{DE}$?*

Solution: Let x represent the length of $\overline{DE}$, so that $100 = \left(\frac{1}{2}\right)(x)(30)$.

This simplifies to $100 = 15x$. Finally, $x = 6\frac{2}{3}$ inches.

3   Example: *Given that DF = 30, DE = 10, and FH = 24, what is the length of $\overline{EG}$?*

Solution: Let's first determine the area of △*DEF*, which is

$\left(\frac{1}{2}\right)(DE)(FH) = \left(\frac{1}{2}\right)(10)(24) = 120$.

Now use the fact that the area of △*DEF* can also be determined by $\left(\frac{1}{2}\right)(DF)(EG)$. Let x represent the length of $\overline{EG}$.

Then $120 = \left(\frac{1}{2}\right)(30)(x)$. Simplify to $120 = 15x$. Finally, $x = 8$.

There is a shortcut you can use for this type of problem. Since the product of a base and its corresponding height is really double the area of the triangle, (DF)(EG) must equal (DE)(FH).
So, (30)(EG) = (10)(24). Then (30)(EG)=240, so EG = 8.

Consider Figure 8.2, showing obtuse $\triangle KLM$, involving its three medians, with the altitude $\overline{KS}$ drawn from point K.

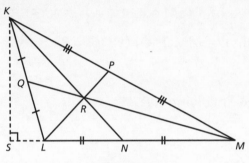

Figure 8.2

If you use $\overline{LM}$ as the base, the area of $\triangle KLM$ is equal to $\left(\dfrac{1}{2}\right)(LM)(KS)$. Since N is the midpoint of $\overline{LM}$, we can replace $\left(\dfrac{1}{2}\right)(LM)$ with LN. This means that the area of $\triangle KLM$ can be written as $(LN)(KS)$. Now let's consider $\triangle KLN$. Using LN as the base, the area of $\triangle KLN$ is $\left(\dfrac{1}{2}\right)(KS)(LN)$. So we are saying that the area of $\triangle KLN$ is exactly one-half the area of $\triangle KLM$. This implies that the area of $\triangle KNM$ is also one-half of the area of $\triangle KLM$. It is logical to say that the area of $\triangle KLN$ added to the area of $\triangle KNM$ is equal to the area of $\triangle KLM$.

The conclusion that we have reached is that a median of a triangle divides it into two triangles, each of which has an area equal to one-half that of the original triangle.

Of course, we could have used the median $\overline{PL}$. In that case, the area of each of $\triangle KPL$ and $\triangle MPL$ is one-half of the area of $\triangle KLM$.

4 **Example:** *Using the median $\overline{QM}$ as a side of each of two triangles, the area of which two triangles is exactly one-half the area of $\triangle KLM$ in Figure 8.2?*

Solution: The median $\overline{QM}$ creates the triangles $\triangle KMQ$ and $\triangle LMQ$, each of which has an area equal to one-half of the area of $\triangle KLM$.

5 **Example:** *Given that the area of $\triangle KLM$ is 84 and that KS = 16, what is the value of MN?*

Solution: The area of $\triangle KLM$ equals $\left(\dfrac{1}{2}\right)(LM)(KS) = (MN)(KS)$. Let x represent the value of *MN*. Then $(x)(16) = 84$, so $x = 5.25$.

Consider Figures 8.3 and 8.4, involving $\triangle TUV$.

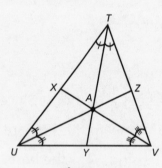
Figure 8.3

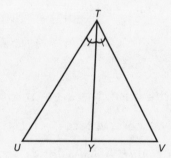
Figure 8.4

In Figure 8.3, all three angle bisectors are drawn; in Figure 8.4, only the angle bisector $\overline{TY}$ is shown. Figure 8.3 shows that all three angle bisectors intersect at a common point *A*.

In Figure 8.4, use your ruler to measure the segments $\overline{TU}$, $\overline{TV}$, $\overline{UY}$, and $\overline{VY}$ in inches. Be as accurate as possible. You will probably get four different numbers that appear completely unrelated. Now calculate the ratios $\dfrac{TU}{TV}$ and $\dfrac{UY}{VY}$.

If these two ratios look about equal, you are on the right path!
Before we draw any conclusions, let's look at Figure 8.5.

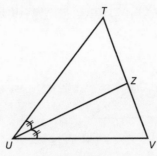

Figure 8.5

Here is the same triangle *TUV* with only the angle bisector $\overline{UZ}$. Again, use your ruler to carefully measure the segments $\overline{TU}$, $\overline{UV}$, $\overline{TZ}$, and $\overline{ZV}$. Calculate the ratios $\dfrac{TU}{UV}$ and $\dfrac{TZ}{ZV}$. Do these answers look equal? They absolutely should!

The conclusion that we have reached is that an angle bisector divides the side to which it is drawn into two parts in the same ratio as the sides forming the original angle. If you want additional reassurance for this conclusion, you could draw just the angle bisector $\overline{XV}$. You should be able to show that $\dfrac{TV}{UV} = \dfrac{TX}{XU}$.

6 | **Example:** | *In Figure 8.4, suppose that TU = 10, TV = 6, and UY = 4. How long is the segment $\overline{YV}$?*

Solution: Let *x* represent the length of $\overline{YV}$. Then $\dfrac{10}{6} = \dfrac{4}{x}$.

The easiest way to solve for *x* is to cross-multiply. We get $10x = 24$, so $x = 2.4$.

7 | **Example:** | *In Figure 8.5, suppose that TU = 10, UV = 12, and ZV = 5. How long is the segment $\overline{TZ}$?*

Solution: Let *x* represent the length of $\overline{TZ}$. Then $\dfrac{10}{12} = \dfrac{x}{5}$.

Cross-multiply to get $12x = 50$, so $x = 4\dfrac{1}{6}$.

Consider Figure 8.6, involving △*WXY*, with its three midsegments.

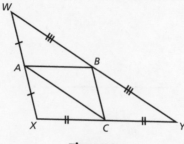

Figure 8.6

SEE LESSON 7 Each midsegment is parallel to the third side, which is the side that is not connected to the midsegment. Also, the length of the midsegment is one-half the length of this third side. Let's redraw this figure as Figure 8.7, with the addition of the altitude of △*WXY* from point *W*.

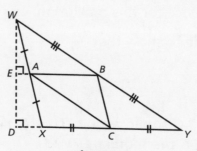

Figure 8.7

SEE LESSON 4 We have extended both $\overline{AB}$ to point *E* and $\overline{XY}$ to point *D*. We know that $\overline{WD}$ is the altitude for △*WXY*. However, ∠*WEB* is also 90°. $\overline{EB}$ and $\overline{DY}$ are already parallel line segments, since they are extensions of $\overline{AB}$ and $\overline{XY}$. If we consider $\overline{WD}$ a transversal, ∠*WEB* and ∠*WDY* are corresponding angles. Thus, these two angles are congruent.

Furthermore, $\overline{WE}$ is the altitude for △*WAB*. Use your ruler to measure the lengths of $\overline{WE}$ and $\overline{ED}$. Don't be surprised if they appear to be equal in length! That is exactly the relationship between those segments. This means that $WE = \left(\dfrac{1}{2}\right)(WD)$.

We also know that $AB = \left(\dfrac{1}{2}\right)(XY)$. We are now going to compare the area of △*WAB* to the area of △*WXY*. Try to guess what relationship exists between these two values. (Hint: Think of a type of horseracing.)

Suppose you are given that $XY = 28$ and $WD = 18$. Then the area of $\triangle WXY$ is $\left(\dfrac{1}{2}\right)(28)(18) = 252$. Based on the given information, we can conclude that $AB = \left(\dfrac{1}{2}\right)(28) = 14$ and that $WE = \left(\dfrac{1}{2}\right)(18) = 9$.

Then the area of $\triangle WAB$ is $\left(\dfrac{1}{2}\right)(14)(9) = 63$.

Now notice that 63 is exactly one-fourth of 252.

Conclusion: For any triangle we can draw its three midsegments. Each of the four smaller triangles formed has an area equal to one-fourth the area of the original triangle.

By the way, the reference to horseracing is that certain types of horses are called "quarter horses." They are known for their fast speeds over short distances, so they are only raced for distances close to one-fourth mile.

Test Yourself!

1. The area of $\triangle PQR$ is 125 square inches. The height of the altitude drawn from point Q is 20 inches. What is the length, in inches, of $\overline{PR}$?

    *Answer:*_____

2. Which one of the following, when drawn in a triangle, creates two smaller triangles with equal areas?

    (A) Altitude   (C) Median

    (B) Angle bisector   (D) Midsegment

3.   **Look at the following figure:**

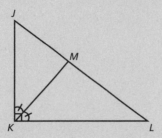

Based only on the symbols shown, which <u>two</u> of the following statements <u>must</u> be true?

(A)   *JM = ML*

(B) $\overline{KM}$ is perpendicular to $\overline{JL}$.

(C)   *m∠JKM = 45°*

(D)   *JK = KL*

(E) $\dfrac{JK}{KL} = \dfrac{JM}{ML}$

4.   **Look at the following figure:**

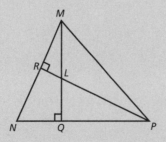

If *NP* = 40, *MN* = 36, and *MQ* = 30, what is the length of $\overline{PR}$?

Answer: _____

Test Yourself! (continued)

5.   In △*STU*, all three midsegments are drawn. Point *V* is the midpoint of $\overline{ST}$, point *W* is the midpoint of $\overline{SU}$, and point *X* is the midpoint of $\overline{TU}$. Which one of the following statements is __not__ necessarily true?

(A)   *VW* = *WX*

(C) $\overline{WX}$ is parallel to $\overline{ST}$.

(B)   *SV* = *VT*

(D) $VX = \left(\dfrac{1}{2}\right)(SU)$

6.   Refer back to the given statements in question 5. If *VW* = 9, how many other line segments __must__ have a length of 9?

(A)   None

(C)   Two

(B)   One

(D)   Three

7.   Look at the following figure, in which *B* is the midpoint of $\overline{ZC}$:

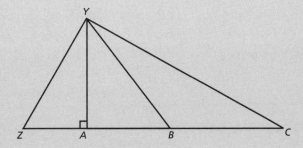

Which __two__ of the following statements __must__ be true?

(A)   *AZ* = *AB*

(B) $\overline{YB}$ is a median of △*YZC*.

(C)   The area of △*YZB* is equal to the area of △*YCB*.

(D)   The area of △*YZA* is equal to the area of △*YAB*.

(E)   *YB* = *BC*

Test Yourself! (continued)

8. Suppose a right triangle is scalene. If you draw all medians, altitudes, and angle bisectors, how many different line segments are drawn? (Do **not** count the original sides of the right triangle.)

    (A)   9   (C)   7

    (B)   8   (D)   6

9. Look at the following equilateral triangle *DEF,* in which *EG = GF.*

    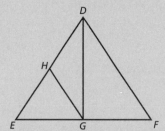

    For any equilateral triangle, a median is also an angle bisector and an altitude. Based on this information, which **two** of the following statements **must** be true?

    (A)   *DH = HE*

    (B)   ∠*EDG* is congruent to ∠*FDG.*

    (C) $\overline{HG}$ is parallel to $\overline{DF}$.

    (D)   △*DGF* is isosceles.

    (E) $\overline{DG}$ is perpendicular to $\overline{EF}$.

10.   **Look at the following figure:**

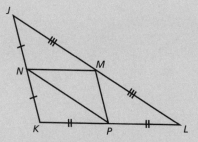

If the combined area of △MNP, △MLP, and △KNP is 43.5 square inches, what is the area, in square inches, of △JKL?

*Answer:*_____

Triangles—Part 4

In this lesson, we will explore properties that are unique to "special triangles," namely, those that are isosceles, right, or both. You should be certain that you know the basic properties of these triangles, as discussed in Lesson 6. All the concepts you learned in previous lessons on triangles will be applicable to these triangles. We use right triangles to, for example, (a) find the height of a tree that is casting a shadow, (b) estimate the angle of elevation for a ladder leaning against a building, and (c) calculate the distance between third base and first base in baseball. The isosceles triangle is used, for example, to (a) identify each face of the Transamerica Pyramid in San Francisco and (b) identify the two parts of a kite formed when drawing its short diagonal.

Your Goal: When you have completed this lesson, you should be able to determine missing lengths of right triangle segments as well as those of isosceles triangle segments. You will also use the area concept that you learned in Lesson 8.

LESSON 9

Triangles—Part 4

We will begin with Figure 9.1, which is a right triangle *ABC,* with altitude $\overline{CD}$.

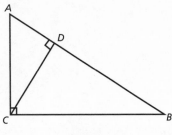

Figure 9.1

You recall that each of $\overline{AC}$ and $\overline{BC}$ is also an altitude for this triangle. As a quick review, let's use Example 1.

1 Example: *Using Figure 9.1, if AC = 9, BC = 12, and AB = 15, what is the length of $\overline{CD}$?*

Solution: The area of $\triangle ABC$ is $\left(\dfrac{1}{2}\right)(AC)(BC) = \left(\dfrac{1}{2}\right)(9)(12) = 54$. Using $\overline{AB}$ as the base and letting *x* represent the length of $\overline{CD}$, we can write $54 = \left(\dfrac{1}{2}\right)(15)(x)$. This equation simplifies to $54 = 7.5x$. So, $x = 7.2$.

The value of AB in Example 1 was actually already known because we knew the values of AC and BC. We are about to find out why this is so.

In any right triangle, the two sides that form the right angle are called **legs**, and the side opposite the 90° angle is called the **hypotenuse**. In any triangle, the largest side is always opposite the largest angle. So, for a right triangle, the largest side will always be the hypotenuse.

It can be shown that for any right triangle, the sum of the squares of its two legs is equal to the square of the hypotenuse. This is called the **Pythagorean theorem**, named after the Greek mathematician Pythagoras. You will almost always see this written as $a^2 + b^2 = c^2$.

Using Figure 9.1, we can state that $(AC)^2 + (BC)^2 = (AB)^2$.
If you look at the given information in Example 1, you can verify that $9^2 + 12^2 = 15^2$.
The left side of this equation is 81 + 144, which equals 225, the value of the right side.

MathFlash!

Never try to "simplify" $(AC)^2 + (BC)^2 = (AB)^2$ into $AC + BC = AB$. This would be just plain wrong!

2 **Example:** *Using Figure 9.1, what is the length of $\overline{AD}$?*

Solution: If we apply the Pythagorean theorem to $\triangle ACD$ and let x represent the length of $\overline{AD}$, the equation starts as $(AD)^2 + (CD)^2 = (AC)^2$.
Then, by substitution, $x^2 + (7.2)^2 = (9)^2$.
Then $x^2 + 51.84 = 81$.
Subtract 51.84 from both sides of the equation to get $x^2 = 29.16$.
In order to complete this problem, we need to take the square root of 29.16 to the nearest hundredth (if necessary).
As it turns out, the answer is exactly 5.4.

3 **Example:**   *Using Figure 9.1, what is the length of $\overline{BD}$?*

Solution: We could use the Pythagorean theorem for $\triangle BCD$, but there is a quicker way!
Since $AB = 15$ and $AD = 5.4$, the length of $\overline{BD}$ is simply $15 - 5.4 = 9.6$.

MathFlash!

If you had chosen to use the Pythagorean theorem for $\triangle BCD$, you could let x = the length of $\overline{BD}$ and use the equation $x^2 + (7.2)^2 = 12^2$. This equation becomes $x^2 + 51.84 = 144$, followed by $x^2 = 92.16$. The last step is to take the positive square root of 92.16 and get 9.6, which is the value of x.

Let's get some more practice using the Pythagorean theorem; then we will return to a figure similar to Figure 9.1. Our accuracy for answers will be to the nearest hundredth.

4 **Example:**   *Look at Figure 9.2.*

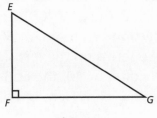

Figure 9.2

Given EF = 13 and EG = 30, what is the value of FG?

Solution: Let x represent *FG*. We know that *EG* is the hypotenuse, so the correct equation is $x^2 + 13^2 = 30^2$.
Simplified, it becomes $x^2 + 169 = 900$, followed by $x^2 = 731$.
So, $x = \sqrt{731} \approx 27.04$.

5 **Example:** *Look at Figure 9.3.*

Figure 9.3

Given HI = 7 and IJ = 12, what is the value of HJ?

Solution: Letting x represent HJ, the equation should be $7^2 + 12^2 = x^2$.
Since $7^2 + 12^2 = 49 + 144$.
Then, $49 + 144 = 193$,
$x = \sqrt{193} \approx 13.89$.

6 **Example:** *In △KLM, the right angle is located at L. If KM = 20 and LM is three times as large as KL, what is the value of KL?*

Solution: Even though no diagram was given, you know that the right angle is at point L. This means that $\overline{KM}$ must be the hypotenuse, because it is the longest side.
Let x represent KL, so that $3x$ can represent LM.
We know that $(KL)^2 + (LM)^2 = (KM)^2$.
By substitution, $x^2 + (3x)^2 = 20^2$. The next step is $x^2 + 9x^2 = 400$.
Be sure you recognized that $(3x)^2 \neq 3x^2$!
Combining like terms on the left side of the equation, we get $10x^2 = 400$. Dividing by 10, the equation will read $x^2 = 40$.
Finally, $x = \sqrt{40} \approx 6.32$. Here is a diagram for this problem:

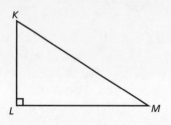

If you want to check the accuracy of this answer, you should determine the value of LM, which is the length of the other leg. From the given information, LM = (3)(6.32) ≈ 18.96. We just need to check that $(6.32)^2 + (18.96)^2 ≈ 20^2$. The left side of this equation is 399.424, and the right side is 400. This is truly a very good approximation.

7 **Example:** *In $\triangle NPQ$, the right angle is located at P. If NP = 16 and NQ is twice as large as PQ, what is the value of NQ?*

Solution: Let's try this one without the diagram. We know that $\overline{NQ}$ is the hypotenuse, so that $(NP)^2 + (PQ)^2 = (NQ)^2$.

Let x represent PQ, and let $2x$ represent NQ.

The equation will be $16^2 + x^2 = (2x)^2$.

Simplified, the equation reads $256 + x^2 = 4x^2$.

Subtracting x^2 from each side, we get $256 = 3x^2$.

This means that $x^2 = \dfrac{256}{3} = 85.\overline{3}$.

Then $x^2 = \sqrt{85.\overline{3}} ≈ 9.24$. But don't stop yet!

The value of x represents PQ, not NQ. Since NQ is represented by $2x$, the approximate value of NQ is $(2)(9.24) = 18.48$.

Check the validity of this answer by verifying that $16^2 + (9.24)^2 ≈ (18.48)^2$. (This is close enough!)

8   **Example:**   *Consider Figure 9.4, a right triangle PQR, with the altitude $\overline{QS}$ to its hypotenuse.*

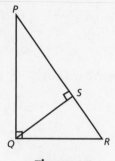

Figure 9.4

Given that PQ = 24 and QR = 10, what is the value of QS?

Solution: Let's first use the Pythagorean theorem for $\triangle PQR$, in order to find PR. Letting x represent PR, we have $10^2 + 24^2 = x^2$.
Simplifying to $676 = x^2$, we find that $x = \sqrt{676} = 26$.
We can determine the area of $\triangle PQR$ by using $\overline{QR}$ as the base and $\overline{PQ}$ as the altitude. The area of $\triangle PQR$ is $\left(\frac{1}{2}\right)(10)(24) = 120$.

Now letting x represent QS, we can assign $\overline{PR}$ as the base and $\overline{QS}$ as the altitude. Then $120 = \left(\frac{1}{2}\right)(26)(x)$, which leads to $13x = 120$.
So, $x = QS \approx 9.23$. ·

9   **Example:**   *Using Figure 9.4, what are the values of PS and SR?*

Solution: Use the Pythagorean theorem for $\triangle PQS$, and let x represent PS, $(9.23)^2 + x^2 = 24^2$.
This equation becomes $85.19 + x^2 = 576$, followed by $x^2 = 490.81$.
So, $x = \sqrt{490.81} \approx 22.15$.
To find the value of SR, simply subtract 22.15 from the length of PR (26) to get 3.85. You can verify this result by using the Pythagorean theorem on $\triangle QRS$. In doing so, you will find that $(9.23)^2 + (3.85)^2 \approx 10^2$.

Look at the values of QS, PS, and PR. If you calculate (PS)(PR), the product is 85.2775. If you calculate (QS)², the value is 85.1929. The difference is only about 0.08, and remember that we did some rounding!

There is a strong relationship among the altitude drawn to the hypotenuse of a right triangle and the two "pieces" of the hypotenuse formed by the intersection of the altitude and hypotenuse.

The square of the height (length of the altitude) drawn to the hypotenuse of a right triangle will equal the product of the two segments of the hypotenuse formed by the altitude. In Figure 9.4, we can state that $(QS^2) = (PS)(SR)$.

10 **Example:**   *Consider Figure 9.5, right triangle TUV with altitude UW.*

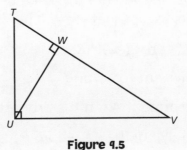

Figure 9.5

Given that TW = 5 and WV = 12, what is the value of UW?

Solution: Let x represent UW. In place of writing $(UW)^2 = (TW)(WV)$, we can write $x^2 = (5)(12) = 60$. Then $x = \sqrt{60} \approx 7.75$.

Let's now return to Figure 9.1. It now has all segment values:

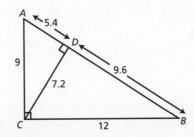

Notice that all values of the line segments have been included. Each of these values has either been given in Examples 1, 2, 3, or has been calculated. In particular, look at the values of *CD*, *AD*, and *DB*. At first glance, there may appear to be nothing special about these numbers. But, based on Example 10, we now know that $(7.2)^2 = (5.4)(9.6)$.

11 **Example:** *Consider Figure 9.6, right triangle XYZ with altitude YA.*

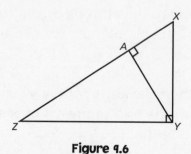

Figure 9.6

Given AZ = 22 and AY = 15, what is the value of XZ?

Solution: Don't panic! We will use the same technique as we used for Example 10, except that we will need to determine *AX* first.

Let's write $(AY)^2 = (AZ)(AX)$.

This is equivalent to writing $\dfrac{AZ}{AY} = \dfrac{AY}{AX}$.

Letting *x* represent *AX* in the proportion, we have $\dfrac{22}{15} = \dfrac{15}{x}$.

Cross-multiplying, the equation reads $22x = 225$, so $x \approx 10.23$. Now, it is easy to see that $XZ = 22 + 10.23 = 32.23$.

MathFlash!

A good warm-up exercise to prepare you for the Drill Exercises would be to see if you can determine the values of ZY and XY. These answers will be provided at the end of the Drill Exercises. Try not to peek!

1. In right triangle *XYZ*, the right angle is at point *X*. If *XZ* = 14 and *XY* = 7, what is the value of *YZ*?

    Answer: _____

2. In right triangle *UVW*, the right angle is at point *V*. If *UV* = 29 and *UW* = 32, what is the value of *VW*?

    Answer: _____

3. In right triangle *RST*, the right angle is at point *S*. *RT* = 24, and *RS* is three times as large as *ST*. What is the value of *ST*?

    Answer: _____

4. In right triangle *NPQ*, the right angle is at point *N*. *NQ* = 30, and *PQ* is twice as large as *NP*. What is the value of *PQ*?

    Answer: _____

5. Look at the following figure:

    Answer: _____

    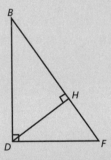

    Given that *BD* = 15 and *BF* = 17, what is the value of *DF*?

6. Referring to the figure in question 5, what is the value of *DH*?

    Answer: _____

7. Look at the following figure: Answer: _____

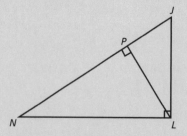

Given that *JP* = 6 and *PN* = 16, what is the value of *PL*?

8. Referring to the figure in question 7,   Answer: _____
    what is the value of *LN*?

9. Look at the following figure: Answer: _____

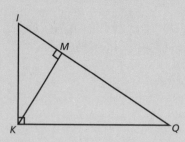

Given that *IM* = 5 and *KM* = 20, what is the value of *QM*?

10. Referring to the figure in question 9,   Answer: _____
    what is the perimeter of △*IKQ*?
    <u>Hint:</u> Be sure to find the values of
    IK and *QK*.

Answers to "Special Note" following Example 11 : *ZY* ≈ 26.63 and *XY* ≈ 18.16

LESSONS 5-9

QUIZ TWO

1. If the measures of two angles of a triangle are 50° and 31°, then this triangle must be _____.

   A acute

   B right

   C obtuse

   D isosceles

2. In △ACE, the measure of ∠C is 90°. If AC = 7 and AE = 18, what is the value of CE to the nearest hundredth?

   A 16.58

   B 17.49

   C 18.42

   D 19.31

3. In △GJL, m∠G = 45°. The measure of ∠L is 15° less than the measure of ∠J. Which one of the following inequalities is correct?

   A GJ < JL < GL

   B GJ < GL < JL

   C JL < GJ < GL

   D JL < GL < GJ

4. In a quadrilateral, two of the sides are 10 inches and 16 inches. The third side is four times as large as the fourth side. If the perimeter is 51 inches, what is the sum of the lengths of the two largest segments?

   A 20 inches

   B 25 inches

   C 30 inches

   D 36 inches

5. Consider the following figure.

   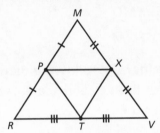

   Which one of the following angles __must__ be congruent to ∠V?

   A ∠PXT

   B ∠R

   C ∠XPT

   D ∠MPX

6. In △BDF, BD = 30, the altitude drawn from point F has a length of 18, and the altitude drawn from point B has a length of 24. What is the value of DF?

   A 22.5

   B 20

   C 14.4

   D 12

7. Consider the following figure.

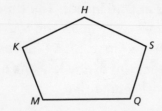

Which one of the following is <u>not</u> an acceptable way to name this polygon?

A MQSHK

B KQMSH

C SHKMQ

D QSHKM

8. Consider the following diagram, in which lines ℓ_1 and ℓ_2 are parallel to each other.

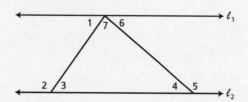

If $m\angle 1 = 63°$ and $m\angle 4 = 48°$, which one of the following inequalities is correct?

A $m\angle 7 < m\angle 3 < m\angle 2$

B $m\angle 1 < m\angle 5 < m\angle 2$

C $m\angle 3 < m\angle 6 < m\angle 5$

D $m\angle 6 < m\angle 1 < m\angle 7$

9. Consider the following right triangle WXY, in which $\overline{XZ}$ is the altitude to $\overline{WY}$.

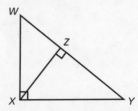

If $WZ = 9$ and $WY = 27$, what is the value of XZ to the nearest hundredth?

A 15.59

B 14.16

C 12.73

D 11.32

10. In the following diagram, $BG = 16$, $BH = 26$, and $\overline{BL}$ is the angle bisector of $\angle GBH$.

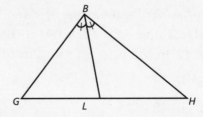

If LH is 8 units longer than GL, what is the value of LH?

A 24.8

B 20.8

C 16.8

D 12.8

Quadrilaterals—Part I

In this lesson, we will explore properties that belong to three different four-sided polygons. Our discussion will include both angles and sides. You are already somewhat familiar with these types of geometric figures from the material in Lesson 5. If you feel that you need to, take a few moments to review the introduction to that lesson.

Your Goal: When you have completed this lesson, you should be able to determine the categories of these three four-sided polygons and understand the basic properties of each category.

LESSON 10

Quadrilaterals—Part 1

Any four-sided polygon is called a **quadrilateral**. Consider Figure 10.1, which is a general quadrilateral with no two congruent sides and with no two congruent angles.

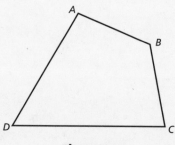

Figure 10.1

The sum of the measures of the angles for any quadrilateral is 360°.

1 **Example:** *Using Figure 10.1, if m∠A = 110°, m∠B = 124°, and m∠C = 72°, what is the measure of ∠D?*

Solution: The sum of the measures of the three given angles is 306°, so $m\angle D = 360° - 306° = 54°$.

2 **Example:** *In a quadrilateral MNPQ, m∠M = 48°, m∠N = 80°, and the measure of ∠P is three times the measure of ∠Q. What is the measure of ∠P?*

Solution: Let x represent $m\angle Q$, so that $3x$ represents $m\angle P$.
Then write $48 + 80 + 3x + x = 360°$.
Simplifying the left side leads to $128 + 4x = 360°$.
$4x = 232°$, so that $x = 58°$. Thus, the measure of $\angle P$ is $(3)(58) = 174°$.

Each angle of any quadrilateral will be less than 180°.

The first specific quadrilateral we will introduce is the **square**. A square is a quadrilateral with four right angles and all equal sides. Opposite sides are also parallel to each other. For this figure, the perimeter is four times the length of any side. Its area is given by the formula $A = s^2$, where s represents the length of one side.

Figure 10.2 shows square *EFGH* with its diagonals $\overline{EG}$ and $\overline{HF}$ intersecting at point *J*.

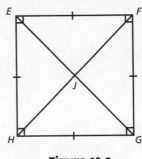

Figure 10.2

Here are some additional properties for the diagonals of a square:

1.   They are congruent.

2.   They are perpendicular bisectors of each other.

3.   They intersect at right angles.

4.   They are angle bisectors wherever they intersect the vertices.

Suppose you want to find the length of a diagonal.
You are given *EF* = 6. $\triangle EFG$ is an isosceles right triangle.
Letting *x* represent *EG*, we can use the Pythagorean theorem to write $6^2 + 6^2 = x^2$. This equation simplifies to $72 = x^2$, so $x = \sqrt{72} \approx 8.49$.

However, you don't have to set up an equation every time for this type of problem. In truth, the length of the diagonal of a square is found by multiplying the length of one side by $\sqrt{2}$. This formula can be written as $d = s\sqrt{2}$.

Now, if you are given the length of the diagonal and want to find the length of a side, just divide by $\sqrt{2}$.

3 **Example:**   *The length of the diagonal of a square is 16. What is the approximate length of one of its sides? (The formula is $s = \dfrac{d}{\sqrt{2}}$.)*

Solution: The length of one side is $\dfrac{16}{\sqrt{2}} \approx 11.31$.

4 **Example:**   *The length of the diagonal of a square is 20. What is the approximate perimeter and area?*

Solution: The length of one side is $\dfrac{20}{\sqrt{2}} \approx 14.14$. Then the perimeter is just $(4)(14.14) = 56.56$, and the area is $(14.14)^2 \approx 199.94$.

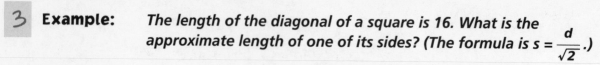

MathFlash!

There is a quick way to find the area of a square if you just want to use the length of the diagonal. (This is your lucky day!) For a square, whose length of a diagonal is known, the area is equal to one-half the square of the diagonal's length.

In Example 4, the area is $\left(\dfrac{1}{2}\right)(20)^2 = \left(\dfrac{1}{2}\right)(400) = 200$.

(199.94, is the result of rounding off when using $\sqrt{2}$.)

The second specific quadrilateral to be introduced is the **rectangle**. A rectangle is a quadrilateral with four right angles, but not necessarily four equal sides. Opposite sides are parallel and equal to each other. A rectangle has two different size lengths for its sides. Traditionally, the longer side is called the **length**, and the shorter side is called the **width**. These two words are used for both the actual segments and their corresponding sizes. The perimeter is found by doubling the length, doubling the width, and then adding these numbers.

The formula is $P = 2l + 2w$, where *l* represents the length, and *w* represents the width. The **area of a rectangle** is the product of the length and width, given by the formula $A = lw$.

Figure 10.3 shows rectangle *KLMN* with its diagonals $\overline{KM}$ and $\overline{LN}$ intersecting at point *P*.

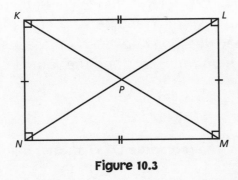

Figure 10.3

Here are some additional properties for the diagonals of a rectangle: (Notice that they were also mentioned as properties for the diagonals of a square.)

1. They are congruent.

2. They bisect each other.

MathFlash!

Be aware that the diagonals in rectangles <u>do not</u> intersect at right angles and <u>do not</u> bisect any of the four angles. Also, if you only know the lengths of the diagonals you do not know enough to find either the perimeter or the area.

5 **Example:**   *In a given rectangle, the length is five inches longer than the width. The perimeter is 88 inches. What is its width?*

Solution: Let *x* represent the width, and let (*x* + 5) represent the length.
Then we can write 88 = 2*x* + 2(*x* + 5).
Removing the parentheses leads to 88 = 2*x* + 2*x* + 10.
This is followed by 88 = 4*x* + 10.
Finally, 78 = 4*x*, which yields *x* = 19.5 inches.

MathFlash!

You can always go the extra step to check this answer. The length would be 19.5 + 5 = 24.5, and we can verify that 88 = (2)(24.5) + (2)(19.5). If the question had asked for the area, you would have had to find both the length and the width.

6 **Example:** *Consider the rectangle QRST shown below:*

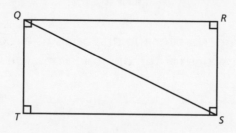

Given that QS = 25 and QT = 10, what is the area of the rectangle?

Solution: We need the value of *ST*, which is the length, so let's use the Pythagorean theorem on △*QST*.
Let *x* represent *ST* to write $x^2 + 10^2 = 25^2$.
Then $x^2 + 100 = 625$, $x^2 = 525$,
and then $x = \sqrt{525} \approx 22.91$.
The area of the rectangle is (10)(22.91) = 229.1.

7 **Example:** *Consider the rectangle UVWX shown below:*

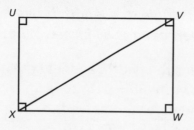

Given that the area is 342 square inches and XW = 36 inches, what is the value of XV?

Solution: We will need the value of *VW* (or *UX*). Use the area formula for a rectangle, and let *x* represent *VW*, 342 = 36*x*. Then *x* = 9.5.
Now, use the Pythagorean theorem on △*VWX*, and let *x* represent *XV*. Write $36^2 + (9.5)^2 = x^2$. This simplifies to $1386.25 = x^2$.
Finally, $x = \sqrt{1386.25} \approx 37.23$ inches.

The third specific quadrilateral to be introduced in this lesson is the **rhombus**. Years ago, there was a dance called the "Twist." You can picture the rhombus as a "twisted" square. All four sides are equal, but the angles at the vertices need not be 90°. Opposite sides are parallel to each other.

Figure 10.4 shows rhombus *YZAB*, with its diagonals $\overline{AY}$ and $\overline{BZ}$ intersecting at point C.

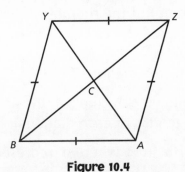

Figure 10.4

If you measure ∠*BAZ* and ∠*BYZ*, you will find them congruent. Likewise, you can verify that *m*∠*AZY* = *m*∠*ABY*. Thus, opposite angles of a rhombus are congruent.

Can you guess the relationship between ∠*YBA* and ∠*BAZ*? By measuring these two angles, you will find that their sum is 180°. ∠*YZA* and ∠*BYZ* are also supplementary angles.

Here is a list of all the properties for the diagonals of a rhombus: (Since Figure 10.4 is drawn to scale, you can verify these statements.)

1. They are perpendicular bisectors of each other.

2. They intersect at right angles.

3. They are angle bisectors wherever they intersect the vertices.

The perimeter of a rhombus is four times the length of any side. (Same as the square.) The area of a rhombus could be determined by multiplying one of its sides by an altitude drawn to that side.

Figures 10.5 and 10.6 are redrawings of Figure 10.4 without the diagonals, each of which contains an altitude to $\overline{AB}$.

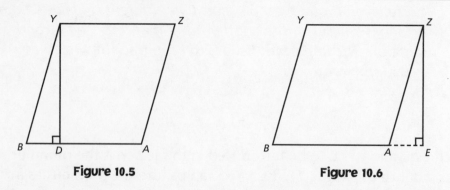

Figure 10.5   Figure 10.6

In Figure 10.5, $\overline{YD}$ is a segment perpendicular to $\overline{AB}$. In Figure 10.6, $\overline{ZE}$ is a segment perpendicular to $\overline{BE}$, which is an extension of $\overline{BA}$. As you would guess, $YD = ZE$. The area of the rhombus can be written as $(BA)(YD)$ or $(BA)(ZE)$.

If the height (length of the altitude) of the rhombus is not given, there is still another way to find the area.

> The key is to know the size of the diagonals.
>
> Similar to the *Math Flash* following Example 4, the area of a rhombus is one-half the product of the length of each diagonal.
>
> The area formula is $A = \left(\dfrac{1}{2}\right)(d_1)(d_2)$, where d_1 represents the length of one diagonal and d_2 represents the length of the other diagonal.

8 **Example:**   *Consider the rhombus FGHJ, as shown below:*

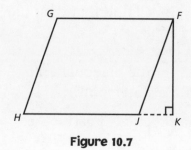

Figure 10.7

$\overline{HJ}$ **is extended to** $\overline{HK}$ **such that** $\overline{KF}$ **is perpendicular to** $\overline{HK}$. **If the area of the rhombus is 1080 and KF = 30, what is the perimeter?**

Solution: Use the area formula, and let x represent HJ.
1080 = 30x. So, $x = 36$. The perimeter is simply $(4)(36) = 144$.

9 **Example:**   *The lengths of the diagonals of a rhombus are 15 and 22. What is the area?*

Solution: The area is $\left(\dfrac{1}{2}\right)(15)(22) = 165$.

10 **Example:**   *The area of a rhombus is 600 square inches. One diagonal is five times as large as the other diagonal. What is the length of the longer diagonal?*

Solution: Let x represent the length of the shorter diagonal, and let $5x$ represent the longer diagonal.

Then $600 = \left(\dfrac{1}{2}\right)(x)(5x)$. Simplify to $600 = 2.5x^2$.

Next divide both sides by 2.5 to get $240 = x^2$.

So, $x = \sqrt{240} \approx 15.49$.
The longer diagonal is $(5)(15.49) = 77.45$ inches.

11 **Example:**   *Consider the rhombus LMNP, as shown below.*

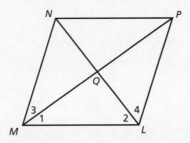

Diagonals PM and LN intersect at point Q. What is the sum of the measures of ∠1 and ∠2?

Solution: ∠*PNM* and ∠*LMN* are supplementary angles, so we can determine that $m\angle1 + m\angle2 + m\angle3 + m\angle4 = 180°$. Since $m\angle1 = m\angle3$ and $m\angle2 = m\angle4$, $m\angle1 + m\angle2 = 90°$.
Another way to do this problem is to recognize that $m\angle MQL = 90°$ and then focus on right triangle *QML*. Its two acute angles must add up to 90°.

12 Example: *Using the diagram in Example 11, suppose the perimeter of the rhombus is 60 and PM = 24. What is the length of $\overline{LN}$?*

Solution: Since the perimeter is 60, each side must be $60 \div 4 = 15$.

Let's use $\triangle PQN$. $PN = 15$, and since there are four right angles at Q, $\overline{PN}$ is the hypotenuse.

Recalling that diagonals of a rhombus bisect each other,

$PQ = \left(\frac{1}{2}\right)(24) = 12$. Let x represent QN.

Then $x^2 + 12^2 = 15^2$. This is simplified to $x^2 + 144 = 225$, followed by $x^2 = 81$.

So $x = \sqrt{81} = 9$. Finally, $LN = (2)(9) = 18$.

MathFlash!

When you identify a quadrilateral by its vertices, the traditional way is to begin with any vertex, then "move" around the quadrilateral one vertex at time, in either direction. Here are four (out of a total of eight) correct ways to name the rhombus in Example 11: (a) LMNP, (b) NPLM, (c) MLPN, (d) PNML Two <u>incorrect</u> ways would be NLPM and MPLN.

Test Yourself!

1. In quadrilateral *ABCD*, $m\angle A = 168°$, $m\angle B = 116°$, and the measure of $\angle C$ is 12° larger than the measure of $\angle D$. What is the measure of $\angle D$?

   Answer: _____

2. Suppose we know that quadrilateral *EFGH* has four congruent sides. Which of the following is the correct conclusion?

   (A) *EFGH* must be a rectangle.

   (B) *EFGH* must be a square.

   (C) *EFGH* must be a square or a rhombus.

   (D) *EFGH* must be a rectangle or a rhombus.

Test Yourself! (continued)

3.   Which one of the following is <u>not</u> a property of a rectangle *IJKL*?

(A)   Opposite sides are congruent.

(B)   The diagonals meet at right angles.

(C)   The diagonals bisect each other.

(D)   The measures of $\angle I$ and $\angle J$ are equal.

4.   In a given rectangle, the length is seven times the width. If the perimeter is 176, what is the length?

Answer: _____

5.   Suppose we know that in quadrilateral *MNPQ*, the diagonals bisect the angles that they intersect. Which of the following is the correct conclusion?

(A)   *MNPQ* must be a rhombus or a square.

(B)   *MNPQ* must be a square.

(C)   *MNPQ* may be a rhombus, a square, or a rectangle.

(D)   *MNPQ* must not be a rhombus, square, or rectangle.

6.   The length of one side of a square is 7.4. To the nearest hundredth, what is the length of one diagonal?

Answer: _____

7.   The area of a rhombus is 99 square inches. If the length of one diagonal is 9 inches, what is the length of the other diagonal?

Answer: _____

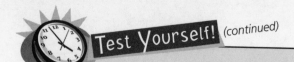 (continued)

8.   The diagonal of a rectangle is 26, and the width is 8. To the nearest hundredth, what is the perimeter?

Answer:_____

9.   Consider the following diagram:

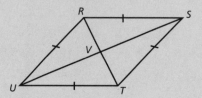

Given that *RU* = 34 and *RT* = 32, what is the area of this figure?

Answer:_____

10.   In a certain quadrilateral, ∠*W* and ∠*B* are opposite angles. The other opposite angles are ∠*Y* and ∠*D*. Which one of the following is <u>not</u> a correct way to name this quadrilateral?

(A)   *WDBY*   (C)   *DYWB*

(B)   *BDWY*   (D)   *YWDB*

11

Quadrilaterals—Part 2

In this lesson, we will explore properties that belong to two new four-sided polygons, also known as quadrilaterals. As in Lesson 10, our discussion will include both angles and sides. Thus far, we have discussed the square, rectangle, and rhombus. It would be correct to say that a square is a particular type of rectangle, since all sides are congruent. It would also be correct to say that a square is a particular type of rhombus, since all angles are congruent.

Your Goal: When you have completed this lesson, you should be able to determine the categories of these new quadrilaterals, understand the basic properties of each category, and be able to compare them to the quadrilaterals you studied in Lesson 10.

LESSON 11

Quadrilaterals—Part 2

The fourth quadrilateral on our agenda is the **parallelogram**. Figure 11.1 illustrates this type of quadrilateral.

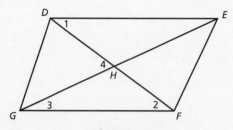

Figure 11.1

Just as we labeled a rhombus as a "twisted" square, we can label a parallelogram as a "twisted" rectangle. The definition of a parallelogram is a quadrilateral in which opposite sides are parallel.

Here is a helpful list of the additional properties of any parallelogram. Since Figure 11.1 is drawn to scale, you can verify these properties by measurement of the sides and angles. (The symbol for the word "congruent" is ≅.)

1. Opposite sides are congruent. Thus, $\overline{DE} \cong \overline{FG}$ and $\overline{DG} \cong \overline{EF}$.

2. Opposite angles are congruent. Thus, $\angle DGF \cong \angle DEF$ and $\angle GDE \cong \angle GFE$.

3. The diagonals bisect each other. Thus, $\overline{DH} \cong \overline{HF}$ and $\overline{EH} \cong \overline{HG}$.

Additionally, we should mention that $\angle DGF$ and $\angle GFE$ are supplementary angles. In any polygon, angles that share a common side are called **consecutive** angles. In Figure 11.1, $\angle FED$ and $\angle EDG$ are also consecutive angles, the sum of whose measures is 180°.

It is equally important to recognize properties that do <u>not</u> belong to the diagonals of a general parallelogram, as shown in Figure 11.1. They do <u>not</u> intersect at right angles, and they do <u>not</u> bisect the angles that they intersect.

Let's use Figure 11.2, which is essentially Figure 11.1 with the addition of two altitudes.

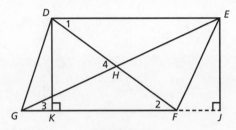

Figure 11.2

The perimeter of a parallelogram is identical to that of a rectangle. Each side is called a **base**, for the altitude drawn perpendicular to it.

In Figure 11.2, $\overline{EJ}$ is the altitude to the base $\overline{GF}$. The perimeter of *DEFG* can be calculated as $(2)(GF) + (2)(EF)$. The area of *DEFG* is found by the product of the length of any base and its corresponding height. Thus, for Figure 11.2, the area is $(GF)(EJ)$.

In the perimeter formula for Figure 11.2, we can substitute DE for GF and DG for EF. Also, any perpendicular segment from $\overline{DE}$ to $\overline{GF}$ can serve as the altitude when using $\overline{GF}$ as the base. Thus, $\overline{DK}$ can be substituted for $\overline{EJ}$.

1 **Example:** *Using Figure 11.1, if the perimeter of DEFG is 34 and GF = 11, what is the value of DG?*

Solution: Let *x* represent the value of *DG*. Then $(2)(11) + (2)(x) = 34$.
The next step is $22 + 2x = 34$, followed by $2x = 12$. Thus, $x = 6$.

2 **Example:** *Using Figure 11.1, with the numbered angles involved, if $m\angle 1 = 38°$ and $m\angle 3 = 23°$, what is the measure of $\angle 4$?*

Solution: Using $\overline{DF}$ as a transversal for the line segments $\overline{DE}$ and $\overline{GF}$, $\angle 1$ and $\angle 2$ are alternate interior angles of parallel line segments. Then $m\angle 2 = 38°$. The easiest way to finish this problem is to recall the meaning of an exterior angle of a triangle and write $m\angle 4 = m\angle 2 + m\angle 3 = 61°$.

3 Example: *Consider parallelogram LMNP, as shown below in Figure 11.3.*

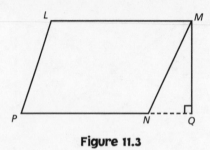

Figure 11.3

If the area of LMNP is 104 and PN is twice as large as MQ, what is the value of PN?

Solution: Let x represent MQ, and let $2x$ represent PN. Then $(x)(2x) = 104$. The next steps are $2x^2 = 104$, $x^2 = 52$, and $x = \sqrt{52} \approx 7.21$, which is the value of MQ. Finally, $PN = (2)(7.21) = 14.42$.

4 Example: *Let's use the information we now know about Figure 11.3. If NQ = 5, what is the perimeter of LMNP?*

Solution: We need to find MN, so let's use the Pythagorean theorem for $\triangle MNQ$. Letting x represent MN, the equation is $(7.21)^2 + 5^2 = x^2$. This equation simplifies to $76.9841 = x^2$, so $x = \sqrt{76.9841} \approx 8.77$. Now the perimeter of $LMNP$ is $(2)(14.42) + (2)(8.77) = 46.38$.

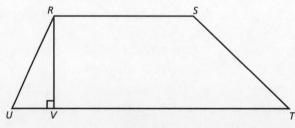

MathFlash!

The square, rectangle, and rhombus are special types of parallelograms.

The fifth quadrilateral for discussion is the **trapezoid**. Figure 11.4 illustrates this type of quadrilateral.

Figure 11.4

A trapezoid is a quadrilateral with <u>exactly</u> two parallel sides, called the **bases**. In Figure 11.4, $\overline{RS}$ and $\overline{UT}$ are the bases. Similar to a parallelogram, $\overline{RV}$ is called the **altitude**. All four sides of trapezoid *RSTU* may have different lengths. The only property that we can verify is that since $\overline{RS}$ is parallel to $\overline{UT}$, $\angle RUT$ and $\angle SRU$ are supplementary angles. The same is true of $\angle STU$ and $\angle RST$.

The **perimeter of a trapezoid** would be found by simply adding the lengths of all four sides. The **area** is found by the formula $A = \left(\dfrac{1}{2}\right)(h)(b_1 + b_2)$, where *h* represents the height, and b_1 and b_2 represent bases. This formula is read as follows: the area equals the product of one-half the height and the sum of the bases.

5 **Example:** *What is the area of a trapezoid in which the height is 5 inches, one base is 6 inches, and the other base is 10 inches?*

Solution: $A = \left(\dfrac{1}{2}\right)(5)(6 + 10) = \left(\dfrac{1}{2}\right)(5)(16) = (2.5)(16) = 40$ square inches.

6 **Example:** *Using Figure 11.4, if the area is 108, RV = 9 and RS = 7, what is the value of UT?*

Solution: Let *x* represent *UT*, which we will denote as b_1.
Then $108 = \left(\dfrac{1}{2}\right)(9)(x + 7)$.
Multiply by 2 to get $216 = (9)(x + 7)$.
This leads to $216 = 9x + 63$. Then $153 = 9x$, so $x = 17$.

7 **Example:** *Consider trapezoid WXYZ shown below as Figure 11.5, with altitude $\overline{WA}$.*

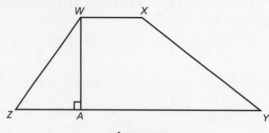

Figure 11.5

Given that WZ = 16, WX = 12, WA = 8, and AY = 25, what is the area of trapezoid WXYZ?

Solution: We need to find the value of *ZA*. Use the Pythagorean theorem for $\triangle WAZ$, and let *x* represent *ZA*, $x^2 + 8^2 = 16^2$.
Simplify this equation to $x^2 = 192$.
Then $x = \sqrt{192} \approx 13.86$.
So *ZY* = 38.86.
The area is $\left(\dfrac{1}{2}\right)(8)(12 + 38.86) = (4)(50.86) = 203.44$.

MathFlash!

In Figure 11.4, RSTV is also a trapezoid. Two of its angles are right angles. Likewise, in Figure 11.5, WXYA is also a trapezoid, in which there are two right angles. When you are computing the area of a trapezoid, be sure that you only apply the factor of $\dfrac{1}{2}$ once. For example, to calculate $\left(\dfrac{1}{2}\right)(6)(18)$, there are three correct ways:

(a) $\left(\dfrac{1}{2}\right)(6) = 3$, then $(3)(18) = 54$.

(b) $\left(\dfrac{1}{2}\right)(18) = 9$, then $(9)(6) = 54$.

(c) $(6)(18) = 108$. Then $\left(\dfrac{1}{2}\right)(108) = 54$.

*Do **not** take one-half of 6 and one-half of 18.*

An **isosceles trapezoid** is defined as one in which the base angles at either base are congruent. Because there are congruent base angles, the nonparallel sides are also congruent. Figure 11.6 illustrates isosceles trapezoid *BCDE*, with altitude $\overline{BF}$.

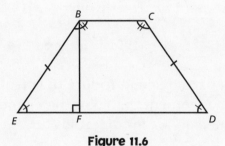

Figure 11.6

For clarity, we have noted that $\angle E \cong \angle D$, $\angle EBC \cong \angle DCB$, and $\overline{BE} \cong \overline{CD}$. Of course its area formula is identical to that of an "ordinary" trapezoid. Let's redraw Figure 11.6 as 11.7, with the diagonals $\overline{BD}$ and $\overline{CE}$ included.

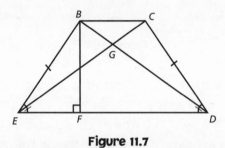

Figure 11.7

Since Figure 11.7 was drawn to scale, use your ruler to measure $\overline{BD}$ and $\overline{CE}$. You should find that they are congruent. Thus, the diagonals of an isosceles trapezoid are congruent.

8 **Example:**   *Using Figure 11.6, if the measure of ∠C is 64° larger than the measure of ∠D, what is the measure of ∠EBC?*

Solution: Let *x* represent the measure in degrees of $\angle D$, and let *x* + 64 represent the measure in degrees of $\angle C$. Since they are supplementary angles, we can write *x* + (*x* + 64) = 180, which becomes 2*x* = 116. So *x* = 58°. Then the measure of $\angle C$ is 58 + 64 = 122°. Since $\angle EBC$ is congruent to $\angle C$, its measure is also 122°.

9 **Example:**   *Using Figure 11.6, if the area of trapezoid BCDE is 1050, BC = 18, and DE = 42, what is the value of BF?*

Solution: Let x represent BF. Then $1050 = \left(\dfrac{1}{2}\right)(x)(18 + 42)$. Since

$\left(\dfrac{1}{2}\right)(18 + 42) = \left(\dfrac{1}{2}\right)(60) = 30$, we have $1050 = 30x$. So, $x = 35$.

Let's return to the general trapezoid and introduce the **median**. The median of a trapezoid is defined as the line segment that joins the midpoints of the two nonparallel sides. Figure 11.8 illustrates median $\overline{ML}$ in trapezoid $HIJK$.

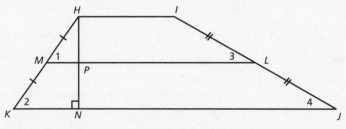

Figure 11.8

Use your protractor to measure $\angle 1$, $\angle 2$, $\angle 3$, and $\angle 4$. You should find that $\angle 1 \cong \angle 2$ and $\angle 3 \cong \angle 4$. This information is sufficient to tell us that $\overline{ML}$ is parallel to each of $\overline{HI}$ and $\overline{JK}$. Of course $HILM$ and $MLJK$ are also trapezoids. Also, it can be shown that the length of the median is equal to the average of the lengths of the two bases.

Mathematically, we can write that $ML = \dfrac{HI + JK}{2}$. If you think back to the area formula for a trapezoid, we can write the area of trapezoid $HIJK$ as the expression $\left(\dfrac{1}{2}\right)(HN)(HI + JK)$. With just a little algebraic juggling, we can also write the area as $(HN)\left(\dfrac{HI + JK}{2}\right)$. This means that the area of trapezoid $HIJK$ can be expressed as the product of the length of the altitude (height) and the length of its median. From the diagram, this is easy to remember. For any trapezoid, draw a vertical line from top to bottom (altitude) and draw a horizontal line across the middle (median). Then just multiply these lengths to get the area!

Incidentally, since M is the midpoint of $\overline{HK}$ and L is the midpoint of $\overline{IJ}$, it should not be surprising that P is also the midpoint of $\overline{HN}$. You can confirm this statement with your ruler.

10 **Example:** *In a given trapezoid, the length of one base is 27, and the length of the median is 21. What is the length of the other base?*

Solution: Let x represent the length of the other base. Then $21 = \dfrac{x + 27}{2}$.

Multiply both sides of the equation by 2 to get $42 = x + 27$.
So, $x = 15$.

11 **Example:** *Suppose the area of HIJK in Figure 11.8 is 42. If HN = 3.5, what is the length of $\overline{ML}$?*

Solution: Let x represent the length of $\overline{ML}$. Even though the lengths of the two bases are not known, we can calculate the length of the median $\overline{ML}$ from the equation $3.5x = 42$. Then $x = 12$.

12 **Example:** *Using Figure 11.8 and the information in the solution of Example 11, suppose that HI = 8. What is the area of trapezoid HILM?*

Solution: Since P is the midpoint of HN, $HP = 1.75$.
Using the original formula for the area of a trapezoid,

we need to calculate $\left(\dfrac{1}{2}\right)(1.75)(8 + 12) = 17.5$.

MathFlash!

If you were now asked to calculate the area of trapezoid MLJK, the easiest method would be to subtract the area of HILM from the area of HIJK. The correct answer would be $42 - 17.5 = 24.5$.

Test Yourself!

1. The perimeter of parallelogram *ACEG* is 76. If *AC* = 25, what is the value of *CE*?

   Answer: _____

2. In parallelogram *HJLN*, the measure of ∠*H* is 44° more than the measure of ∠*J*. What is the measure of ∠*N*?

   Answer: _____

3. Consider parallelogram *QSUW*, as shown below.

   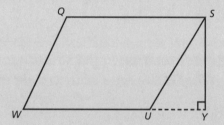

   If *US* = 15, *WU* = 30, and *UY* = 5, what is the area of *QSUW* to the nearest whole number?

   Answer: _____

4. In trapezoid *BDFH*, the length of one base is three times the size of the length of the other base.

   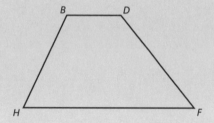

   If the altitude is 10 and the area is 150, what is the length of the longer base?

   Answer: _____

Test Yourself! (continued)

5.   Which one of the following is <u>not</u> necessarily true about an isosceles trapezoid labeled *JLNP where* $\overline{PN}$ *is the longer base*?

   (A)   *PL = JN*

   (B) $\overline{PL}$ is perpendicular to $\overline{JN}$.

   (C) $\angle PJL \cong \angle JLN$

   (D) $m \angle PJL + m \angle JPN = 180°$

6.   In a certain trapezoid, the length of the median is five times as large as the altitude. If the area is 425 square inches, how many inches long is the altitude?

   Answer: _____

7.   In isosceles trapezoid *RTVX*, *RT* and *VX* are the bases. If *RX* = 27, *RT* = 17, and the length of the median is 32, what is the perimeter of *RTVX*?

   Answer: _____

8.   Which one of the following statements is true?

   (A)   A rhombus is a particular type of rectangle.

   (B)   A parallelogram is a particular type of trapezoid.

   (C)   A trapezoid is a particular type of rhombus.

   (D)   A rectangle is a particular type of parallelogram.

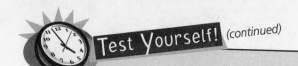
Test Yourself! (continued)

9.   Consider trapezoid *AHMP*, as shown below.

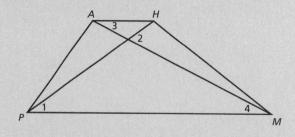

If $m\angle1 = 24°$ and $m\angle3 = 18°$, what is the sum of the measures of $\angle1$ and $\angle2$?

(A)   66°   (C)   48°

(B)   60°   (D)   42°

10.   Consider parallelogram *WXYZ*, for which $\overline{XT}$ is an altitude.

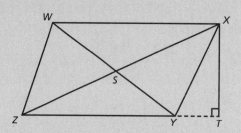

Given that *SZ* = 15, *ZY* = 19, and *YT* = 5, what is the area of *WXYZ*?

Answer: _____

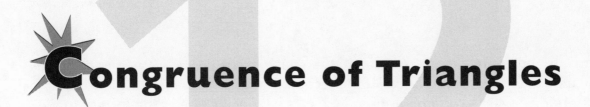

Congruence of Triangles

In this lesson, we will explore the meaning of congruence between two triangles. You have already studied congruence between two angles and between two sides in such figures as triangles, squares, and rectangles. In a physical sense, the concept of congruence between any two one-dimensional or two-dimensional geometric figures is that we can place one figure on top of another in such a way that they coincide exactly. For three-dimensional figures, two congruent figures simply look like duplicates. Some examples of congruence (or near congruence) in everyday life are (a) apples in a bin at the supermarket, (b) a box of paper clips, (c) pages in a notebook, and (d) a collection of quarters.

Your Goal: When you have completed this lesson, you should be able to determine the minimum requirements to establish congruence between two given triangles.

LESSON
12

Congruence of Triangles

If two triangles are already congruent, then each of the angles in the first triangle will match up to a specific angle in the second triangle. Also, each of the sides in the first triangle will be congruent, or match up, to a specific side in the second triangle. Consider Figures 12.1 and 12.2, shown below.

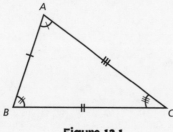

Figure 12.1

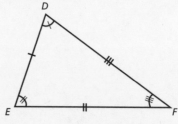

Figure 12.2

As shown by the tick marks, $\angle A \cong \angle D$, $\angle B \cong \angle E$, $\angle C \cong \angle F$, $\overline{AB} \cong \overline{DE}$, $\overline{AC} \cong \overline{DF}$, and $\overline{BC} \cong \overline{EF}$. To show that the triangles are congruent, we write $\triangle ABC \cong \triangle DEF$.

MathFlash!

When we name a single triangle, any arrangement of the letters used for the vertices is allowed. Thus, in Figure 12.1, the triangle can be named in ways other than $\triangle ABC$, such as $\triangle BCA$ and $\triangle CBA$. However, when identifying a congruence between two triangles, the specific order of the letters used is very important. The order of the letters tells us which angles and sides are actually congruent. For example, if $\triangle GHI \cong \triangle KLJ$, then we conclude that $\angle G \cong \angle K$ and that $\overline{HI} \cong \overline{JL}$.

Let's discuss the minimum requirements to conclude that two triangles are actually congruent. These requirements apply to any type of triangle.

One minimum requirement is called **Side-Side-Side**. This means that if each of the three sides in one triangle can be matched with each of the three sides in the second triangle, the triangles are automatically congruent.
Figures 12.3 and 12.4 illustrate this situation.

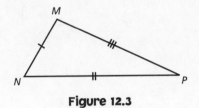
Figure 12.3

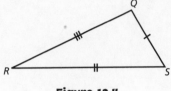
Figure 12.4

Based on the tick marks, $\triangle MNP \cong \triangle QSR$. Although they are not marked as congruent, we are able to conclude that $\angle M \cong \angle Q$, $\angle N \cong \angle S$, and $\angle P \cong \angle R$.

Notice that if we were just given Figures 12.3 and 12.4, we could still establish which angles are congruent. Just locate any two congruent sides; the respective angles opposite these sides <u>must</u> be congruent. For example, since $\overline{NP} \cong \overline{RS}$, we would know that $\angle M \cong \angle Q$.

MathFlash!

Do not be disturbed by the fact that the triangles don't appear congruent as you are viewing them. If Figure 12.4 could be physically "turned over" (like a pancake), it would fit on top of Figure 12.3. Looking at Figures 12.5 and 12.6, we can see another example of how $\triangle MNP \cong \triangle QSR$ could appear.

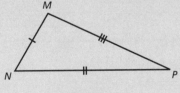
Figure 12.5

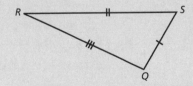
Figure 12.6

Another minimum requirement for congruence is called **Side-Angle-Side**. This means that each of two sides in one triangle is matched with each of two sides in a second triangle. Also, the included angle of the first triangle is congruent to the included angle of the second triangle. It is vital that the congruent angles lie between the pairs of congruent sides.

Figures 12.7 and 12.8 illustrate this situation.

Based on the tick marks, $\overline{TV} \cong \overline{WX}$, $\overline{UV} \cong \overline{XY}$, and $\angle V \cong \angle X$.
This is sufficient information to conclude that $\triangle TUV \cong \triangle WYX$.
Now, we can also state that $\overline{UT} \cong \overline{WY}$, $\angle T \cong \angle W$, and $\angle U \cong \angle Y$.

MathFlash!

The importance of the location of the pair of congruent angles <u>*between*</u> *the two pairs of congruent sides cannot be stressed enough. Figures 12.9 and 12.10 illustrate two triangles.*

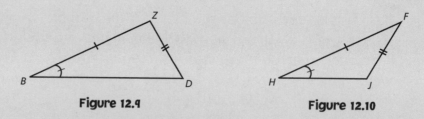

*The tick marks show that $\overline{ZB} \cong \overline{FH}$, $\overline{ZD} \cong \overline{FJ}$, and $\angle B \cong \angle H$. Yet these triangles are not congruent. This situation is sometimes called the **Side-Side-Angle fallacy**.*

A third minimum requirement is called **Angle-Side-Angle**. This means that each of two angles in one triangle can be matched with each of two angles in a second triangle, plus the included side of the first triangle is congruent to the included side of the second triangle.

Figures 12.11 and 12.12 illustrate this situation.

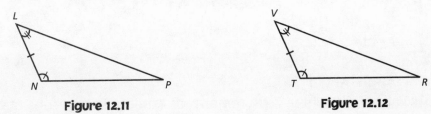

Figure 12.11   **Figure 12.12**

The tick marks show that $\angle L \cong \angle V$, $\angle N \cong \angle T$, and $\overline{LN} \cong \overline{VT}$. This is sufficient information to conclude that $\triangle LNP \cong \triangle VTR$.
Consequently, we can further state that $\angle P \cong \angle R$, $\overline{LP} \cong \overline{VR}$, and $\overline{NP} \cong \overline{TR}$.

A fourth (and final) minimum requirement is called **Side-Angle-Angle**. (Some textbooks use the phrase "Angle-Angle-Side.") This means that each of two angles in one triangle can be matched with each of two angles in a second triangle, plus a side not included between these two angles of the first triangle must match a side not included between these two angles of the second triangle.

Figures 12.13 and 12.14 illustrate this situation.

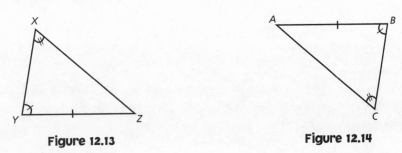

Figure 12.13   **Figure 12.14**

The tick marks show that $\overline{YZ} \cong \overline{AB}$, $\angle Y \cong \angle B$, and $\angle X \cong \angle C$. As in the first three stated requirements, this information is sufficient to claim that $\triangle XYZ \cong \triangle CBA$. In addition, we can also state that $\angle Z \cong \angle A$, $\overline{XY} \cong \overline{BC}$, and $\overline{XZ} \cong \overline{AC}$.

MathFlash!

When identifying two triangles as being congruent by Side-Angle-Angle, we must be certain that the angles opposite the sides not included are congruent. In Figures 12.13 and 12.14, if $\overline{YZ}$ had been marked as congruent to $\overline{AC}$ instead of $\overline{AB}$, we could not conclude that the triangles are necessarily congruent.
Remember that whenever two triangles are actually congruent, a pair of congruent sides must <u>always</u> lie opposite a pair of congruent angles.

In summary, we have shown that there are four different sets of minimum requirements to allow us to conclude that two triangles are congruent:

> Side-Side-Side (SSS)
> Side-Angle-Side (SAS)
> Angle-Side-Angle (ASA)
> Side-Angle-Angle (SAA)

Many textbooks will refer to these requirements as "theorems" or "postulates" and use the abbreviations that you see in the parentheses.

> In Figures 12.9 and 12.10, you have seen why Side-Side-Angle does not guarantee congruence of triangles. Can you think of another instance in which three parts of one triangle are congruent to three parts of a second triangle, yet the triangles may <u>not</u> be congruent? The answer is Angle-Angle-Angle (AAA).

Figures 12.15 and 12.16 illustrate this situation.

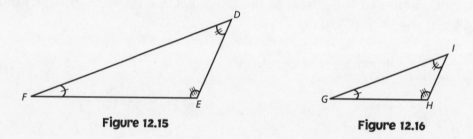

Figure 12.15   **Figure 12.16**

The tick marks show that $\angle D \cong \angle I$, $\angle F \cong \angle G$, and $\angle E \cong \angle H$. However, these triangles are definitely not congruent. You can see that the corresponding sides of $\triangle DEF$ are larger than those of $\triangle IGH$.

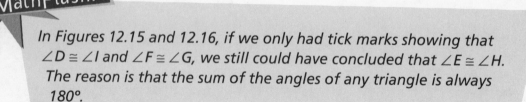

MathFlash!

In Figures 12.15 and 12.16, if we only had tick marks showing that $\angle D \cong \angle I$ and $\angle F \cong \angle G$, we still could have concluded that $\angle E \cong \angle H$. The reason is that the sum of the angles of any triangle is always 180°.

Let's now consider some "special" triangles, in order to determine the minimum requirements in order to establish congruence.

Suppose we have two equilateral triangles, as shown in Figures 12.17 and 12.18.

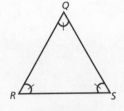

Figure 12.17

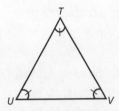

Figure 12.18

Of course, we know that the measure of each angle of both triangles must be 60°. If no other equivalences are given, would this be sufficient to state that the triangles are congruent? Hopefully, you answered no. But, if we can determine that any side of △*QRS* is congruent to any side of △*TUV*, then we would have a solid basis to say that the two triangles are congruent. The reason is Side-Angle-Angle or Angle-Side-Angle.

Now consider two isosceles triangles, as shown in Figures 12.19 and 12.20.

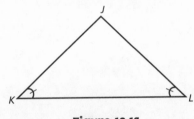

Figure 12.19

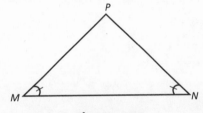

Figure 12.20

Based on the tick marks, we know that $\angle K \cong \angle L \cong \angle M \cong \angle N$. We can certainly conclude that $\angle J \cong \angle P$. This would <u>not</u> be enough information to establish congruence between the triangles. If we knew that *KL = MN*, then the triangles would be congruent by Angle-Side-Angle. Suppose we were given that *JK = PM*. Hopefully, you can spot that the triangles would be congruent by Side-Angle-Angle. Incidentally, $\angle K$ and $\angle L$ are called **base angles** of the isosceles △*JKL*, and $\angle J$ is called the **vertex** angle.

Let's consider one more special triangle, namely, the **right triangle**. Look at Figures 12.21 and 12.22.

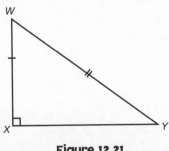

Figure 12.21

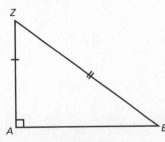

Figure 12.22

At first glance, based on the tick marks, it appears that we have a Side-Side-Angle match, which does not guarantee congruence between the triangles.
But the Pythagorean theorem comes to the rescue!
For Figure 12.21, $(WX)^2 + (XY)^2 = (WY)^2$.

Likewise, for Figure 12.22, $(AZ)^2 + (AB)^2 = (BZ)^2$.
Since $WX = AZ$ and $WY = BZ$, we only have to substitute to see that $XY = AB$.
Now, you can state that $\triangle WXY \cong \triangle ZAB$ by Side-Angle-Side.
(We are using the right angle at both vertex X and vertex A.)

1 **Example:** *Given that $\triangle CED \cong \triangle BGF$, which side is congruent to $\overline{CD}$?*

Solution: Simply match the location of the letters from the first triangle to the second triangle. The answer is $\overline{BF}$.

2 **Example:** *Consider triangles HIJ and KLM, with tick marks as shown.*

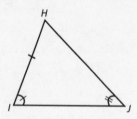
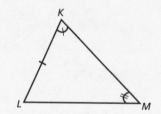

Which one of the following is the correct conclusion?

(A) *The triangles are congruent by Side-Angle-Angle.*

(B) *The triangles are congruent by Side-Angle-Side.*

(C) *The triangles are each isosceles.*

(D) *The triangles are not necessarily congruent.*

Solution: Answer choice (A) is correct. The triangles are congruent by Side-Angle-Angle. We do not have enough information to determine if the two triangles are isosceles.

3 **Example:** *Return to Example 2. Which side of △KLM is congruent to $\overline{IJ}$?*

Solution: Since $\overline{IJ}$ is opposite $\angle H$, and $\angle H \cong \angle L$, the side opposite $\angle L$, which is $\overline{KM}$, must be congruent to $\overline{IJ}$.

4 **Example:** *PQR and STU are each isosceles right triangles, with m∠P = m∠S = 90°. If △PQR ≅ △SUT, which one of the following is <u>false</u>?*

(A) $\angle R \cong \angle U$

(B) $m\angle T = 45°$

(C) $PR = RQ$

(D) *The triangles are congruent by Side-Angle-Side.*

Solution: Answer choice (C) is correct. $\overline{RQ}$ is the hypotenuse of △PQR, so it must be the largest side of that triangle. Each acute angle of these two triangles must have a measure of 45°, and the triangles are congruent by Side-Angle-Side. Technically, they are also congruent by Side-Angle-Angle.

Note that since we can show that $QR = TU$ by using the Pythagorean theorem, even Side-Side-Side would be acceptable for demonstrating that the two triangles are congruent.

5 **Example:**   *Consider the following two triangles.*

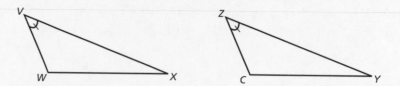

Based on the tick marks, which of the following groups of information could be added in order to guarantee that the triangles are congruent?

(A) *VX = YZ and WX = CY*

(B) *∠X ≅ ∠Y*

(C) *Both triangles are obtuse.*

(D) *∠W ≅ ∠C and VW = CZ*

Solution: Choice (D) is the correct answer. The triangles would be congruent by Angle-Side-Angle.

Answer choice (A) is wrong, since we would only have a Side-Side-Angle relationship.

Answer choice (B) is wrong, since we would only have two pairs of congruent angles.

Answer choice (C) is wrong, since two obtuse triangles need not be congruent. In fact, they may not have any other congruent parts, except for the given tick marks.

MathFlash!

In order to establish that two triangles are congruent, at least one pair of sides must be shown to be congruent. If only pairs of angles can be shown to be congruent, there is no guarantee that the triangles will be congruent.

Test Yourself!

1. If △DGK ≅ △FEJ, which one of the following <u>must</u> be true?

   (A)   ∠G ≅ ∠J

   (B) $\overline{DK} ≅ \overline{JF}$

   (C)   They are both acute triangles.

   (D)   ∠K ≅ ∠F

2. △HLN and △MPR are each isosceles triangles, with ∠H ≅ ∠M. What is the minimum number of pairs of sides that must be congruent in order to conclude that △HLN ≅ △MPR?

   (A)   One

   (B)   Two

   (C)   Three

   (D)   The triangles cannot be congruent.

3. Look at the following triangles.

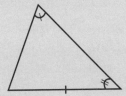 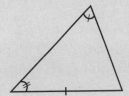

   These two triangles are congruent by the minimum requirement
   _____-_____-_____.

4.   Which one of the following pairs of triangles can <u>never</u> be congruent?

   (A)   Two right triangles

   (B)   An obtuse triangle and an acute triangle

   (C)   An acute triangle and an equilateral triangle

   (D)   An isosceles triangle and a right triangle

5.   Which one of the following does <u>not</u> guarantee congruence between two triangles?

   (A)   Side-Side-Side   (C)   Side-Angle-Angle

   (B)   Side-Side-Angle   (D)   Side-Angle-Side

6.   It is known that $\triangle NST \cong \triangle ZXW$, and that $m\angle N = 90°$. Which one of the following statements must be <u>false</u>?

   (A) $m\angle X = 90°$ (C) $\triangle NST$ is obtuse.

   (B) $m\angle W < 90°$ (D) $NT = WZ$

7.   Suppose two sides of one triangle are congruent to two sides of a second triangle. Which <u>one</u> of the following would each guarantee that the triangles are congruent?

   (A)   Any pair of congruent angles

   (B)   They are both right triangles.

   (C)   They are both isosceles triangles.

   (D)   Congruent angles that are included between the congruent sides

8. If △*AEJ* is isosceles, with *AE = EJ*, then ∠*E* is called a _____ angle, and ∠*J* is called a _____ angle.

9. Consider the following triangles.

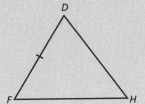 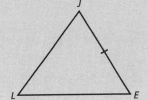

Based on the tick marks, which one of the following could be added in order to guarantee that △*DFH* ≅ △*EJL* ?

(A)   *FH = JL*

(B)   ∠*F* ≅ ∠*E* and ∠*H* ≅ ∠*L*

(C)   ∠*D* ≅ ∠*E* and ∠*F* ≅ ∠*J*

(D)   *DH = EL* and ∠*F* ≅ ∠*J*

Congruence of Quadrilaterals

13

In this lesson, we will explore the meaning of congruence between two quadrilaterals. The general rule is that for each angle in one quadrilateral, there must be a corresponding congruent angle in a second quadrilateral. This same statement must also be true for each of the sides.

Your Goal: When you have completed this lesson, you should be able to determine the requirements needed to establish congruence between specific types of quadrilaterals.

LESSON 13

Congruence of Quadrilaterals

Let's begin with the general case. Consider Figures 13.1 and 13.2, shown below.

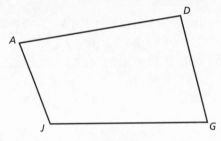

Figure 13.1

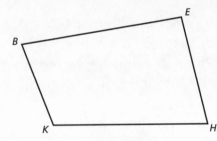

Figure 13.2

If we know that quadrilateral *ADGJ* is congruent to quadrilateral *BEHK*, then each of the following congruencies must hold: ∠*A* ≅ ∠*B*, ∠*D* ≅ ∠*E*, ∠*G* ≅ ∠*H*, ∠*J* ≅ ∠*K*, $\overline{AD} \cong \overline{BE}$, $\overline{DG} \cong \overline{EH}$, $\overline{GJ} \cong \overline{HK}$, and $\overline{AJ} \cong \overline{BK}$.

Let's study the familiar quadrilaterals that you have seen.

Suppose we are given two squares. The only minimum requirement to establish that they are congruent is that a side of one square is congruent to a side of the other square. All angles are 90°, so they are automatically congruent.

Now consider the situation for two given rectangles. Similar to squares, we already know that all angles measure 90°. If only the length of the first rectangle is congruent to the length of the second rectangle, there would not be a guarantee of congruence between the rectangles, as shown in Figures 13.3 and 13.4 on the next page.

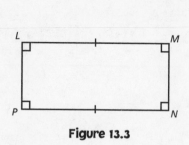

Figure 13.3

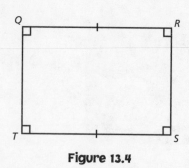

Figure 13.4

Based on the tick marks, you can see that *LP < QT*. These two rectangles are not congruent. Thus, we would need to know that <u>both</u> the length and width of the first rectangle are congruent, respectively, to the length and width of the second rectangle.

Now, let's consider the rhombus. To establish congruence between two rhombi (plural of *rhombus*), we need a congruence between a side of one rhombus and a side of a second rhombus. In addition, how many pairs of congruent angles are necessary to establish this? Just one pair is needed.

Look at Figures 13.5 and 13.6, shown below.

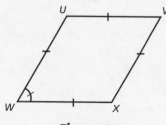

Figure 13.5

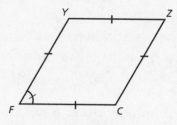

Figure 13.6

All sides are congruent for both rhombi, and $\angle W \cong \angle F$. We know that $\angle W \cong \angle V$ and $\angle F \cong \angle Z$. This means that $\angle V \cong \angle Z$. Also, $\angle W$ and $\angle X$ are supplementary angles; so are $\angle F$ and $\angle C$. This implies that $\angle X \cong \angle Y$ and $\angle U \cong \angle Y$.

If we know that a pair of sides and a single pair of angles are congruent between two rhombi, then the rhombi are congruent.

Now let's look at the parallelogram. To understand the minimum requirements for congruence between two parallelograms, we should view a parallelogram as a "mixture" of rectangle and rhombus.

Using the properties of a rectangle, we know that we must have congruence of both length and width for the two given parallelograms.
Using the language for the sides of a parallelogram, we'll state that corresponding pairs of bases must be congruent. Just one pair of corresponding angles must be congruent. The logic for this is identical to the explanation used for the rhombi.

Figures 13.7 and 13.8, shown below, illustrate these minimum requirements for congruence between two parallelograms.

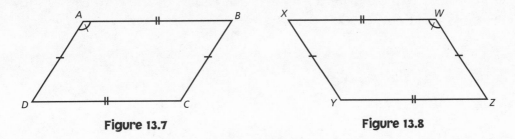

Figure 13.7   Figure 13.8

Based on the tick marks, we can also conclude that $\angle C \cong \angle Y$, and $\angle B \cong \angle D \cong \angle X \cong Z$. Thus, parallelogram *ABCD* is congruent to parallelogram *WXYZ*. Note that another way in which we could express this congruence is $ABCD \cong YZWX$.

The last type of quadrilateral we will investigate is the trapezoid. Consider two congruent trapezoids, *EFGH* and *STUV*, as shown below in Figures 13.9 and 13.10.

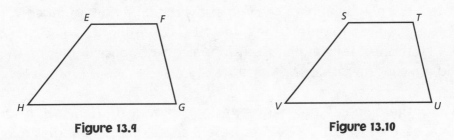

Figure 13.9   Figure 13.10

How many congruent pairs of sides do you think would be a minimum requirement to establish congruence for these trapezoids? Since a trapezoid may have four different lengths for its sides, the correct answer would be four. Remember that the only physical requirement for a trapezoid is that it must have four sides, exactly two of which are parallel to each other.

In this case, $\overline{EF}$ is parallel to $\overline{GH}$, and $\overline{ST}$ is parallel to $\overline{UV}$. Now let's focus on the measures of the angles. Assume that we are given $\angle H \cong \angle V$. Is there another congruence that must be true? $\angle E$ is supplementary to $\angle H$ and $\angle V$ is supplementary to $\angle S$. These relationships imply that $\angle E \cong \angle S$.

Is knowing that $\angle H \cong \angle V$ and $\angle E \cong \angle S$ sufficient to determine the measures of the remaining angles? We can state that $\angle F$ and $\angle G$ are supplementary, as are $\angle T$ and $\angle U$. Unfortunately, there are many pairs of angle measures that add up to 180°, so we cannot be sure that any two of $\angle F$, $\angle G$, $\angle T$, and $\angle U$ are congruent. It seems that besides knowing that $\angle H \cong \angle V$, we must also determine another pair of congruent angles such as $\angle F \cong \angle T$ or $\angle G \cong \angle U$.

If we know that $\angle H \cong \angle V$ and $\angle G \cong \angle U$, we have sufficient information to conclude that $\angle E \cong \angle S$ and $\angle F \cong \angle T$. Thus, for a trapezoid, we need a minimum of two pairs of congruent angles in order to conclude that all four pairs are congruent. The initial two pairs of angles <u>must</u> share the same base.

Finally, let's consider two congruent isosceles trapezoids *IJKL* and *NPQR*, as shown below in Figures 13.11 and 13.12.

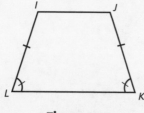

Figure 13.11

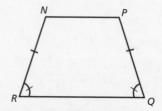

Figure 13.12

Just focusing on the sides, we would need three pairs of congruencies in order to declare that trapezoid *IJKL* $\cong$ trapezoid *NPQR*. We need to know that $\overline{IJ} \cong \overline{NP}$, $\overline{KL} \cong \overline{QR}$, and either one of $\overline{IL}$ or $\overline{JK}$ are congruent to either one of $\overline{NR}$ or $\overline{PQ}$.

With respect to only the angles, can you guess the minimum number of pairs of congruencies required to determine that *IJKL* is congruent to *NPQR*? If you answered "just one," take a giant step forward! Since base angles of an isosceles trapezoid are congruent, $\angle L \cong \angle K$ and $\angle R \cong \angle Q$.

If any two of these angles are congruent, then all four of them are congruent. In addition, each of $\angle I$, $\angle J$, $\angle N$, $\angle P$ represent supplementary angles to any of $\angle L$, $\angle K$, $\angle R$, $\angle Q$. Consequently, all four of these angles would also be congruent. As a numerical example, assume that $m\angle L = m\angle R = 70°$. Then $m\angle K = m\angle Q = 70°$. In addition, it must follow that $m\angle I = m\angle J = m\angle N = m\angle P = 110°$.

1. Consider the following congruent quadrilaterals.

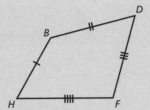 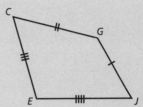

Based on the tick marks, which one of the following represents the correct congruence?

(A) $BDFH \cong GJCE$

(C) $HFDB \cong JECG$

(B) $DBHF \cong GCJE$

(D) $FBDH \cong JECG$

2. Consider the following two figures.

If all angles of each figure are 90°, which additional information will guarantee that *MNPQ* is congruent to *UTSR*?

(A) $\overline{MN} \cong \overline{TU}$ and $\angle Q \cong \angle R$

(C) $\angle N \cong \angle T$ and $\angle P \cong \angle S$

(B) $\overline{MN} \cong \overline{TU}$ and $\overline{PQ} \cong \overline{RS}$

(D) $\overline{PQ} \cong \overline{RS}$ and $\overline{MQ} \cong \overline{RU}$

3. What is the minimum requirement for two squares to be congruent?

(A)   They are automatically congruent.

(B)   One side of the first square is congruent to one side of the second square.

(C)   One side and one angle of the first square are congruent, respectively, to one side and one angle of the second square.

(D)   All four sides of the first square are congruent, respectively, to all four sides of the second square.

4. If quadrilateral *VWXA* is congruent to quadrilateral *ZBFD*, which one of the following is **not** correct?

   (A) $\overline{XA} \cong \overline{DF}$ (C) $\overline{WA} \cong \overline{BF}$

   (B) $\angle W \cong \angle B$ (D) $\angle DZB \cong \angle WVA$

5. In trapezoids *CFHK* and *MPTW*, $\angle C \cong \angle M$, $\angle F \cong \angle P$, $\angle H \cong \angle T$, and $\angle K \cong \angle W$. Which one of the following is the correct conclusion?

   (A) The trapezoids must be congruent.

   (B) If $\overline{CF} \cong \overline{MP}$, then the trapezoids must be congruent.

   (C) If all four pairs of corresponding sides are congruent, the trapezoids must be congruent.

   (D) Even if all four pairs of corresponding sides are congruent, it is impossible for the trapezoids to be congruent.

6. Which of the following would guarantee that rhombi *GJLQ* and *ERSV* are congruent?

   (A) $\overline{GJ} \cong \overline{ER}$ and $\angle L \cong \angle S$

   (B) $\overline{GQ} \cong \overline{EV}$ and $\angle G \cong \angle L$

   (C) $\overline{LJ} \cong \overline{SR}$ and $\overline{LQ} \cong \overline{SV}$

   (D) The rhombi are already congruent.

7. Consider the following isosceles trapezoids.

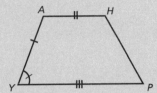 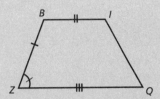

Based on the tick marks, what additional information is required in order to conclude that *AHPY* is congruent to *BIQZ*?

(A) $\angle A \cong \angle B$

(B) $\overline{HP} \cong \overline{IQ}$

(C) $\angle P \cong \angle Q$

(D) The isosceles trapezoids are already congruent.

8. For parallelograms *CDFJ* and *KNRT*, $\overline{CD} \cong \overline{KN}$ and $\overline{FD} \cong \overline{RN}$. To guarantee that the parallelograms are congruent, what is the minimum number of pairs of angles that must be congruent?

(A) One   (C) Three

(B) Two   (D) Four

QUIZ THREE

1. Consider triangles *ACE* and *BDF*, shown below.

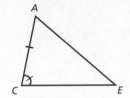 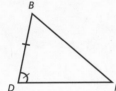

Based on the given tick marks, which one of the following congruencies would <u>not</u> guarantee that △*ACE* ≅ △*BDF*?

A $\overline{AE} \cong \overline{BF}$

B $\overline{CE} \cong \overline{DF}$

C $\angle A \cong \angle B$

D $\angle E \cong \angle F$

2. What is the <u>minimum</u> requirement for rectangle *HKMR* to be congruent to rectangle *STAB*?

A $HK = ST$ and $KM = TA$

B $RM = BA$

C $KM = TA$ and $HR = SB$

D The measure of every angle is 90°.

3. Consider triangles *GJL* and *NQP*, shown below.

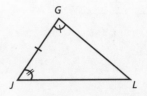 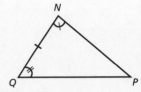

Based on the tick marks, △*GJL* ≅ △*NQP* by which of the following?

A Angle-Angle-Side

B Side-Side-Side

C Angle-Side-Angle

D Angle-Angle-Angle

4. Consider parallelogram *EFGH*, with altitude *FJ*, as shown below.

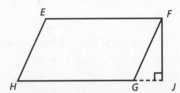

The area of *EFGH* is 261. If *HG* is three times as large as *FJ*, what is the value of *HG* to the nearest hundredth?

A 9.33

B 14.73

C 20.59

D 27.99

5. Consider the rhombus *KPRT*, shown below.

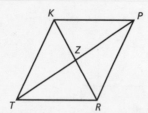

If the perimeter of *KPRT* is 84 and *PZ* = 16, approximately how much larger is the value of *PT* than that of *KR*?

A 4.8

B 9.6

C 14.4

D 19.2

6. What is the area of a trapezoid in which the two bases are 34 inches and 16 inches, and the height is 13 inches?

A 162.5 square inches

B 325 square inches

C 487.5 square inches

D 650 square inches

7. Isosceles right triangle *UVW* is <u>not</u> congruent to isosceles right triangle *XYZ*. The right angles are located at points *V* and *Y*, respectively. *VW* > *YZ*. Which one of the following <u>must</u> be true?

A $m\angle U \neq m\angle X$

B *XY* > *UV*

C $m\angle Y \neq m\angle V$

D *UW* > *XZ*

8. The area of a square is 46.24 square inches. What is the approximate length of a diagonal?

A 15.26 inches

B 12.44 inches

C 9.62 inches

D 6.80 inches

9. In a given trapezoid, the area is 500. If the median is eight units longer than the shorter base, and the longer base is 48 units, what is the height?

A 15.5

B 12.5

C 10.5

D 7.5

10. Consider the isosceles trapezoids *ZYXW* and *DCBA*, shown below. The bases are $\overline{ZY}$, $\overline{WX}$, $\overline{DC}$, and $\overline{AB}$.

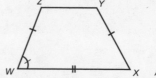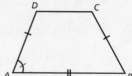

What additional information is necessary in order to conclude that these trapezoids are congruent?

A $m\angle X \cong m\angle B$

B $\overline{ZY} \cong \overline{DC}$

C $m\angle Y \cong m\angle C$

D No additional information is required.

Similarity of Triangles

In this lesson, we will explore the meaning of similarity between two triangles. The concept of similarity for any two geometric figures is that they have the same shape, but one figure is larger than the other figure. You are already familiar with two-dimensional and three-dimensional objects that are similar in everyday life, such as (a) balloons at a fair, (b) mailing envelopes sold in a store, and (c) a scale model of a building and the actual building.

Your Goal: When you have completed this lesson, you should be able to determine the requirements needed to establish similarity between two triangles.

LESSON 14

Similarity of Triangles

Let's begin with the definition of **proportion**. A proportion is simply an equation of two equal ratios. The word **ratio** is equivalent to the word **fraction**. Here are a few examples of proportions.

(a) $\dfrac{1}{2} = \dfrac{4}{8}$ (b) $\dfrac{30 \text{ miles}}{1 \text{ hour}} = \dfrac{90 \text{ miles}}{3 \text{ hours}}$ (c) $\dfrac{\frac{2}{3}}{10} = \dfrac{1}{15}$ (d) $\dfrac{2.4 \text{ inches}}{3.0 \text{ inches}} = \dfrac{4 \text{ feet}}{5 \text{ feet}}$

Two geometric figures are **similar** if the corresponding pairs of angles are congruent and the corresponding pairs of sides are in proportion. We can consider the minimum requirements for concluding that two triangles are similar.

One minimum requirement is called **Angle-Angle**. This means that two angles of one triangle are congruent to two angles of a second triangle.
Once we have two congruent pairs of angles between two triangles, the third pair of angles is automatically congruent.
This is based on the fact that the sum of the angles of any triangle is 180°.

Figures 14.1 and 14.2 illustrate this situation.

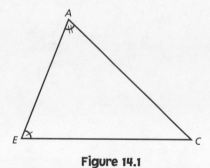

Figure 14.1

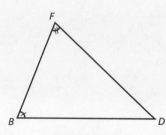

Figure 14.2

Based on the tick marks, △*AEC* is similar to △*FBD*. The symbol for "similar" is "~."
Thus, we can write △*AEC* ~ △*FBD*.

Once the triangles are shown to be similar, we can state that the corresponding sides are in proportion. Mathematically, this information can be written as $\dfrac{AE}{FB} = \dfrac{AC}{FD} = \dfrac{EC}{BD}$.

Remember that corresponding sides must lie opposite congruent angles. (There is no symbol to show similarity for the sides of two triangles.)

Another minimum requirement is called **Side-Side-Side**. This means that the ratio of each pair of corresponding sides of the two triangles forms a proportion. Since there are no tick marks to show similarity among the three pairs of sides, a number must be assigned to each of these sides.

Figures 14.3 and 14.4 illustrate this situation.

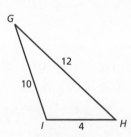

Figure 14.3

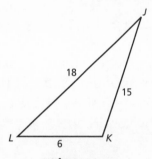

Figure 14.4

We can see that $\dfrac{4}{6} = \dfrac{10}{15} = \dfrac{12}{18}$.

All three pairs of sides <u>must be</u> reducible to the same number. In this case, all three fractions reduce to $\dfrac{2}{3}$. Now, since we can declare that $\triangle GIH \sim \triangle JKL$, we have the following angle congruencies: $\angle G \cong \angle J$, $\angle I \cong \angle K$, and $\angle H \cong \angle L$.

MathFlash!

The easiest way to see if two triangles are similar, given the lengths of all sides, is to determine the ratio of the smallest numbers for each triangle. Then determine the ratio for the largest numbers for each triangle. If this ratio is not the same, the proportion does <u>not</u> exist, and the triangles are not similar.

If this ratio is the same, you must still check the ratio of the remaining two numbers. If this ratio equals the ratio of the other two pairs of numbers, the triangles are similar.

Look at the following two triangles, as illustrated by Figures 14.5 and 14.6.

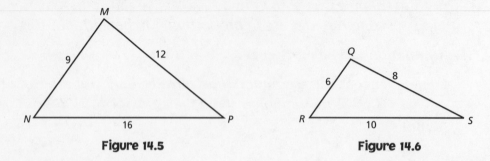

Figure 14.5   Figure 14.6

These triangles certainly do not appear similar, and they are not! But suppose you quickly checked the ratio $\frac{MP}{QR}$, which has a reduced value of 2. Then you checked the ratio $\frac{NP}{QS}$, which is also worth 2. This comparison would be invalid because $\overline{MP}$ is <u>not</u> the shortest side of $\triangle MNP$, <u>but</u> $\overline{QR}$ is the shortest side of $\triangle QRS$. Even if you missed this error, notice that the ratio $\frac{MN}{RS}$ equals $\frac{9}{10}$, not 2.

A third minimum requirement is called **Side-Angle-Side**. This means that the ratio of one pair of corresponding sides of the two triangles must equal the ratio of a second pair of corresponding sides of the two triangles. Also, the included angles of each triangle must be congruent.

Consider Figures 14.7 and 14.8, shown below.

Figure 14.7   Figure 14.8

By inspection, $\frac{9}{3} = \frac{24}{8}$, which reduces to 3. Also, the included angles $\angle U$ and $\angle W$ are congruent. This information is enough to see that $\triangle TUV \sim \triangle XWY$. Therefore, we can also conclude that $\frac{TV}{XY} = 3$. Also, $\angle T \cong \angle X$ and $\angle V \cong \angle Y$.

Using the given information in Figures 14.7 and 14.8, suppose you knew that XY = 6. How would you find the length of TV?

The solution is to let x represent TV and then solve $\frac{x}{6} = 3$. The answer is x = 18.

As a summary, we have shown three different sets of minimum requirements that allow us to conclude that two triangles are similar, namely, Angle-Angle (AA), Side-Side-Side (SSS), and Side-Angle-Side (SAS).

Note that we need not worry about Angle-Side-Angle (ASA) or Side-Angle-Angle (SAA), since two pairs of congruent angles alone would guarantee similarity. Also, one pair of sides <u>cannot</u> form a proportion!

A Side-Side-Angle comparison does not produce a similarity between two triangles. Consider Figures 14.9 and 14.10, shown below.

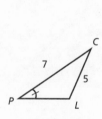

Figure 14.9

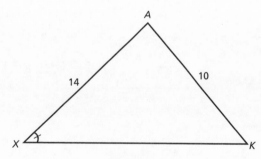

Figure 14.10

You can see that $\frac{CP}{AX} = \frac{CL}{AK} = \frac{1}{2}$, and that $\angle X \cong \angle P$. However, unless you have been sleeping throughout this lesson, these two triangles are definitely not similar! You can see that $\frac{PL}{XK} \neq \frac{1}{2}$. So, now we know that Side-Side-Angle for triangles doesn't work for congruence or similarity.

1 **Example:** *If $\triangle BCF \sim \triangle GLM$, and $\frac{BC}{GL} = \frac{3}{5}$, which other ratio has a value of $\frac{3}{5}$?*

Solution: There are actually two answers. Both $\frac{CF}{LM}$ and $\frac{BF}{GM}$ have a value of $\frac{3}{5}$. (Notice that a diagram is not necessary.)

2 Example: *Given that △DHQ ~ △NPS and that $\frac{PS}{HQ} = \frac{7}{4}$, if DH = 56, what is the value of NP?*

Solution: Let x represent the value of NP. From the given information, we know that $\frac{NP}{DH} = \frac{7}{4}$. By substitution, $\frac{x}{56} = \frac{7}{4}$. Cross-multiply to get $4x = 392$. Thus, $x = 98$.

3 Example: *Return to the information in Example 2. What is the value of $\frac{DQ}{NS}$?*

Solution: Since $\frac{NS}{DQ} = \frac{7}{4}$, then $\frac{DQ}{NS} = \frac{1}{\frac{7}{4}} = \frac{4}{7}$. Incidentally, $\frac{4}{7}$ is called the reciprocal of $\frac{7}{4}$.

4 Example: *Consider triangles FJP and XTU, as shown below.*

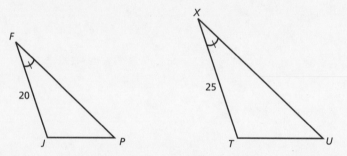

Which one of the following would guarantee that these two triangles are <u>not</u> similar?

(A) $\angle P \cong \angle U$

(B) $\frac{JP}{TU} = \frac{5}{6}$

(C) $JP < TU$

(D) $\overline{FP}$ is the longest side of △FJP.

Solution: The correct answer is (B). Since $\overline{JP}$ and $\overline{TU}$ would represent corresponding sides, the ratio $\frac{JP}{TU}$ must equal $\frac{20}{25} = \frac{4}{5}$. Otherwise, the triangles cannot be similar. Answer choice (A) would actually guarantee that the triangles are similar. No conclusion can be made based on answer choices (C) or (D).

5 **Example:** *In △YZB, m∠Z = 65° and m∠B = 85°. If △YZB ~ △DGR, which one of the following is completely correct?*

(A) *m∠G = 65° and m∠D = 40°*

(B) *m∠Y = 30° and m∠R = 65°*

(C) $\dfrac{YZ}{DG} = \dfrac{65}{85}$ *and m∠R = 85°*

(D) $\dfrac{YB}{DR} = \dfrac{ZB}{GR}$ *and m∠G = 65°*

Solution: The correct answer is (D). The proportion shows the correct ratio of sides of the two triangles. Also, ∠G must be congruent to ∠Z. Answer choice (A) is wrong because the measure of ∠D should be 30°. Answer choice (B) is wrong because the measure of ∠R should be 85°. Answer choice (C) is wrong because the ratio of the corresponding sides is not related to the measures of the angles.

6 **Example:** *Consider similar triangles HKM and NPQ, as shown below.*

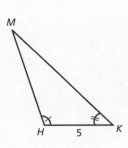
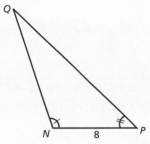

If QP is 18 units larger than MK, what is the length of $\overline{QP}$?

Solution: Let x represent MK, and let $x + 18$ represent QP. Then $\dfrac{5}{8} = \dfrac{x}{x + 18}$. Cross-multiply to get $5x + 90 = 8x$. This leads to $90 = 3x$, which results in $x = 30$, which is MK. Finally, $QP = 48$.

Let's now consider "special" triangles, in terms of similarity.

If we are given any two equilateral triangles that are not congruent, they <u>must</u> be similar. This is because each angle of any equilateral triangle is 60°.

Suppose we have two isosceles triangles, as shown in Figures 14.11 and 14.12.

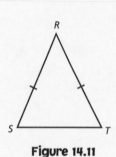

Figure 14.11

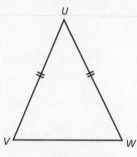

Figure 14.12

When we look at just the sides, we already can determine that $\dfrac{RS}{UV} = \dfrac{RT}{UW}$.

However, to be sure they are similar, we would need to know that the ratio $\dfrac{ST}{VW}$ is equal to $\dfrac{RS}{UV}$ (or to $\dfrac{RT}{UW}$).

We can claim that $\triangle RST \sim \triangle UVW$ if just one pair of angles are equal! To illustrate this last answer, suppose we are given that $m\angle R = m\angle U = 30°$. This is enough to determine that each of $\angle S$, $\angle T$, $\angle V$, and $\angle W$ has a measure of 75°, which guarantees similarity.

Had we known that $m\angle S = m\angle V = 75°$, we would know that both $\angle T$ and $\angle W$ have a measure of 75°, and that both $\angle R$ and $\angle U$ have a measure of 30°.

Now we direct our attention to right triangles.
Look at Figures 14.13 and 14.14, shown below.

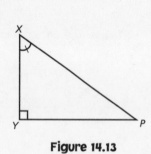

Figure 14.13

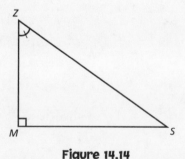

Figure 14.14

Based on the tick marks for the non-right angles, we can claim that $\triangle XYP \sim \triangle ZMS$. If we were not given the information that $\angle X$ is congruent to $\angle Z$, at least how many pairs of equal ratios (proportions) would we need to be certain that these triangles are similar? Hopefully, you said two. The reason is due to our well-known friend Pythagoras. The Pythagorean theorem would be used to find the length of the third sides, and they would represent the same ratio.

To illustrate this, let $XY = 5$, $YP = 12$, $ZM = 15$, and $MS = 36$.

You can easily see that $\dfrac{XY}{ZM} = \dfrac{YP}{MS} = \dfrac{1}{3}$.

Let x represent XP. Then $5^2 + 12^2 = x^2$, which leads to $169 = x^2$. Thus, $x = \sqrt{169} = 13$.

Now let x represent ZS. Then $15^2 + 36^2 = x^2$, which leads to $1521 = x^2$.

Thus, $x = \sqrt{1521} = 39$. Sure enough, $\dfrac{13}{39} = \dfrac{1}{3}$.

MathFlash!

Unfortunately, when we use the Pythagorean theorem, we often don't get whole numbers. Sometimes we get square roots such as the $\sqrt{90}$. Even if XP and ZS had been square root numbers, the good news is that their ratio would still have been $\dfrac{1}{3}$.

7 **Example:**   *Triangles CAK and TRP are shown below.*

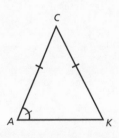 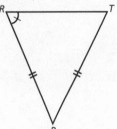

Based on the tick marks, which <u>two</u> of the following are correct?

(A) $\triangle CAK \sim \triangle RPT$ (D) $\triangle KAC \sim \triangle RPT$

(B) $\triangle CKA \sim \triangle PRT$ (E) $\triangle ACK \sim \triangle PRT$

(C) $\triangle AKC \sim \triangle RTP$

Solution: The correct answers are (B) and (C). Remember that the order of the letters indicating the vertices dictates the actual correspondence of the measures of the angles.
Answer choice (B) shows us that $\angle C \cong \angle P$, $\angle K \cong \angle R$, and $\angle A \cong \angle T$. Since both triangles are isosceles, with base angles of $\angle A$, $\angle K$, $\angle R$, and $\angle T$, we can match up either one of $\angle A$ or $\angle K$ with either one of $\angle R$ or $\angle T$. Of course, $\angle C$ must be paired with $\angle P$.
Answer choice (C) is also consistent with the correct correspondences between each pair of angles.

8 **Example:**   **△DFW ~ △EIU and both triangles are isosceles. Given that ∠D and ∠E are the vertex angles, DF = 10, EU = 5, and FW = 6, which one of the following is <u>completely</u> correct?**

(A) **IU = 6 and ∠E ≅ ∠D**

(B) **∠W ≅ ∠E and EI = 5**

(C) **∠W is larger than ∠D and IU = 3.**

(D) **DW = 6 and ∠U ≅ ∠F**

Solution: The correct answer is (C). Recall that within one triangle, the largest angle is always opposite the largest side. ∠W is opposite a side with a length of 10 and ∠D is opposite a side with a length of 6.

Also, *IU* must equal 3, based on solving the proportion $\dfrac{10}{5} = \dfrac{6}{IU}$.

Answer choice (A) is wrong because *IU* does not equal 6.
Answer choice (B) is wrong. ∠W must actually be larger than ∠E, since ∠E and ∠D are congruent.
Answer choice (D) is wrong because *DW* = 10.

Example 8 was not an easy question. Here is a diagram to help you to understand the solution.

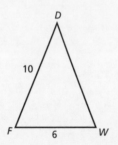
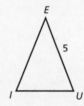

9 **Example:** *Right triangles JLN and QMR are shown below.*

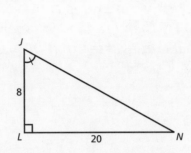
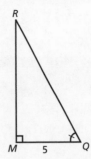

What is the value of MR?

Solution: Based on the tick marks, these triangles must be similar.

Let x represent *MR* and write $\dfrac{8}{5} = \dfrac{20}{x}$.

Then cross-multiply to get $8x = 100$. Finally, $x = 12.5$.

10 **Example:** $\triangle ABC \sim \triangle DEF$, with a right angle at vertices *B* and *E*, as shown below.

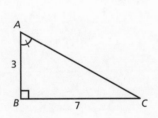
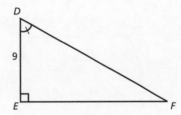

What is the value of DF?

Solution: One way to find *DF* is to first determine *AC*.
It looks like the Pythagorean theorem is visiting us again!
Letting x represent *AC*, we can write $3^2 + 7^2 = x^2$.

By now, you could probably do the next few steps running on empty, but here they are anyway. Start with $9 + 49 = x^2$, followed by $58 = x^2$, and then $x = \sqrt{58} \approx 7.62$.

We are not done yet. To find *DF*, just use an appropriate proportion. Using the information that $AC = 7.62$, and letting x represent *DF*, we can write $\dfrac{3}{9} = \dfrac{7.62}{x}$. This leads to $3x = 68.58$. Finally, $x = 22.86$.

MathFlash!

Another approach you could use in Example 10 is to use a proportion to determine EF. Then use the Pythagorean theorem with the numbers in △DEF. You will see that DF ≈ 22.86.

Test Yourself!

1. Which one of the following is an <u>incorrect</u> proportion?

   (A) $\dfrac{\frac{1}{2}}{8} = \dfrac{3}{48}$

   (C) $\dfrac{7}{77} = \dfrac{0.04}{0.4}$

   (B) $\dfrac{\$3.60}{\$6.00} = \dfrac{6 \text{ hours}}{10 \text{ hours}}$

   (D) $\dfrac{18 \text{ feet}}{8 \text{ feet}} = \dfrac{90 \text{ miles}}{40 \text{ miles}}$

2. Look at the following two triangles.

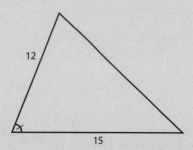
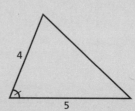

   By what reason are these triangles similar?

   (A) SAS

   (C) SSS

   (B) AA

   (D) SAA

3. If △GIK ~ △LJH, which one of the following is a correct proportion?

   (A) $\dfrac{GI}{LJ} = \dfrac{IK}{LH}$

   (C) $\dfrac{LH}{GK} = \dfrac{JH}{GI}$

   (B) $\dfrac{GK}{HJ} = \dfrac{IG}{LJ}$

   (D) $\dfrac{HJ}{IK} = \dfrac{LH}{GK}$

Test Yourself! (continued)

4. Given that $\triangle MNP \sim \triangle RSQ$, $\dfrac{MN}{RS} = \dfrac{10}{7}$, and $NP = 25$, what is the value of QS?

Answer: _____

5. Referring to the information in question 4, if MP is 9 units larger than RQ, what is the value of RQ?

Answer: _____

6. If $\triangle TUV \sim \triangle XWY$ and $m\angle U = m\angle W = 90°$, which one of the following statements is **not** necessarily true?

(A) $TV > UV$ (C) $XW = WY$

(B) $\dfrac{TU}{XW} = \dfrac{UV}{WY}$ (D) $m\angle V - m\angle T = m\angle Y - m\angle X$

7. Consider triangles ACE and GIK, as shown below.

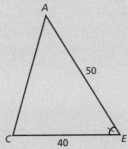 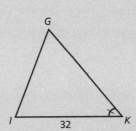

Which one of the following would guarantee that these triangles are **not** similar?

(A) $\dfrac{AC}{GI} = \dfrac{CE}{IK}$

(B) $GK < CE$

(C) $\angle A \cong \angle G$

(D) $\overline{CE}$ is the shortest side of $\triangle ACE$.

Test Yourself! *(continued)*

8.   If $\triangle MPR \sim \triangle NLQ$, $m\angle M = 15°$, $m\angle L = 115°$, and $\dfrac{MP}{NL} = \dfrac{4}{3}$, which one of the following is <u>impossible</u>?

   (A) $\angle P$ is the largest angle of $\triangle MPR$.

   (B) $\dfrac{LQ}{PR} = \dfrac{3}{4}$

   (C) $\angle Q$ is the smallest angle of $\triangle NLQ$.

   (D) $PR < 3$

9.   Consider two isosceles triangles as shown below. These two triangles are similar.

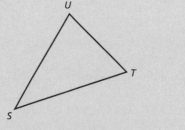
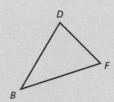

   The base angles of $\triangle SUT$ are $\angle U$ and $\angle T$. Which one of the following would guarantee that the base angles of $\triangle BFD$ are $\angle D$ and $\angle F$?

   (A) $\angle S \cong \angle F$

   (B)   The perimeter of $\triangle SUT$ is twice the perimeter of $\triangle BFD$.

   (C)   Each triangle is obtuse.

   (D) $\triangle STU \sim \triangle BFD$

10. $\triangle XYZ \sim \triangle HKM$, with a right angle at vertices Y and K, as shown below. What is the value of KM?

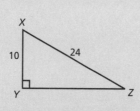
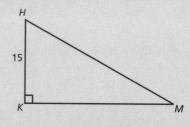

Answer: _____

Similarity of Quadrilaterals

In this lesson, we will explore the meaning of similarity between two quadrilaterals. Just as we found out in Lesson 13, which dealt with congruence of quadrilaterals, corresponding angles of the two quadrilaterals must be congruent. Since we are looking for similarity, any two corresponding pairs of sides must form a proportion.

Your Goal: When you have completed this lesson, you should be able to determine the requirements needed to establish similarity between specific types of quadrilaterals.

Similarity of Quadrilaterals

Let's begin with the general case. Consider Figures 15.1 and 15.2, shown below.

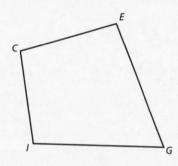

Figure 15.1

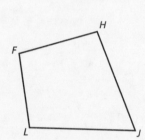

Figure 15.2

If we know that quadrilateral *CEGI* is similar to quadrilateral *FHJL*, then each of the following congruencies must hold: $\angle C \cong \angle F$, $\angle E \cong \angle H$, $\angle G \cong \angle J$, and $\angle I \cong \angle L$.

We also have the following equivalent ratios: $\dfrac{CE}{FH}$, $\dfrac{EG}{HJ}$, $\dfrac{GI}{JL}$, and $\dfrac{IC}{LF}$.
If each of these ratios were inverted, they would also be equivalent.

Suppose we are given two **squares**. The only minimum requirement to establish that they are similar is that there is none! This is because all angles are 90° and therefore equal. Also all of the corresponding sides are in proportion.

Suppose a side of the first square is 8 and a side of the second square is 16. Then the ratio between any side of the first square to any side of the second square must be $\dfrac{8}{16} = \dfrac{1}{2}$.

Suppose we have two **rectangles**. The measure of each angle is 90°. We need to know the length and width of each rectangle in order to determine whether the rectangles are similar. Consider Figures 15.3 and 15.4, as shown below.

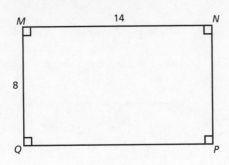

Figure 15.3

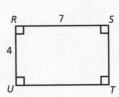

Figure 15.4

These two rectangles are similar because $\dfrac{14}{7} = \dfrac{8}{4} = \dfrac{2}{1}$.

Now lets look at the **rhombus**. To verify similarity between two rhombi, we recognize that with respect to the sides, we use the same reasoning that we used for squares. How about the minimum requirement for angle congruence? Only one pair of congruent angles is required to know that the given rhombi are similar.

Look at Figures 15.5 and 15.6, shown below.

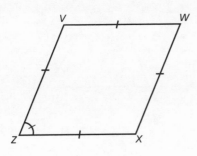

Figure 15.5

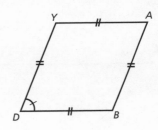

Figure 15.6

These are two rhombi of different sizes for which $\angle Z \cong \angle D$. Since opposite angles of a rhombus are congruent, we really have $\angle Z \cong \angle W \cong \angle D \cong \angle A$.
Now $\angle Z$ and $\angle X$ are supplementary angles, as are $\angle D$ and $\angle B$. This implies that $\angle X \cong \angle B$.
Using this reasoning, it is easy to show that $\angle V \cong \angle Y$.

Our conclusion is that two rhombi are similar if a single pair of angles can be shown to be congruent.

Let's continue on to the **parallelogram**. Figures 15.7 and 15.8 illustrate two parallelograms that appear similar.

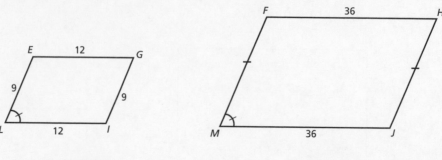

Figure 15.7   Figure 15.8

Based on the numerical values given we need to be certain that *EGIL ~ FHJM*. (~ means "similar to".) We require that $\frac{EG}{FH} = \frac{EL}{FM} = \frac{3}{1}$. This implies that *FM* = 27. The tick marks reveal that $\angle L \cong \angle M$. As with the discussion on the rhombi, this is sufficient information to claim that all corresponding pairs of angles are congruent. So, to show similarity, we need a proportion with two different pairs of corresponding sides and a congruence of one pair of corresponding angles.

Now, let's look at the **trapezoid**. Consider Figures 15.9 and 15.10, as shown below.

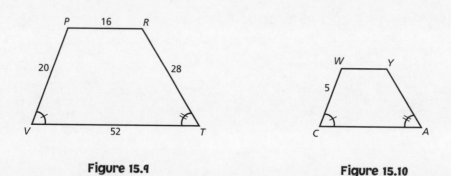

Figure 15.9   Figure 15.10

Based on the tick marks for the lower base angles, $\angle V \cong \angle C$ and $\angle T \cong \angle A$. In terms of the measures of the angles, this would be enough to state that *PRTV ~ WYAC*.

The reason is that $\angle P$ and $\angle W$ are supplementary to $\angle V$ and $\angle C$, respectively. Knowing that $\angle V \cong \angle C$ implies that $\angle P \cong \angle W$. The same argument can be used to explain why $\angle R \cong \angle Y$.

If you need numerical values, assume that m∠V = 70° and m∠T = 50°. Immediately, m∠C = 70° and m∠A = 50°. From all the knowledge you have accumulated on trapezoids, it should be easy to claim that m∠P = m∠W = 110° and that m∠R = m∠Y = 130°.

Concerning the sides, there must be a constant ratio between all corresponding pairs. Since $\frac{PV}{WC} = \frac{20}{5} = 4$, we would need a value of 4 for the other corresponding ratios. Then $WY = 4$, $YA = 7$, and $AC = 13$ in order to claim that the trapezoids are similar.

Before we end this lesson, let's look at isosceles trapezoids, as shown below in Figures 15.11 and 15.12.

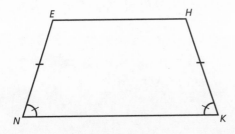

Figure 15.11

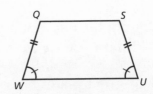

Figure 15.12

Based on the tick marks, since $\angle N \cong \angle K \cong \angle W \cong \angle U$, we can easily show that $\angle E \cong \angle H \cong \angle Q \cong \angle S$. Having just one pair of congruent angles is enough to show that all four pairs of angles are congruent.

Can you guess the minimum number of pairs of sides that must represent the same ratio? If all the angles are congruent in pairs, the isosceles trapezoids will be similar if one pair of corresponding bases is in the same ratio as one pair of corresponding congruent sides.

Test Yourself!

1. Consider the following similar quadrilaterals.

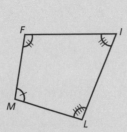 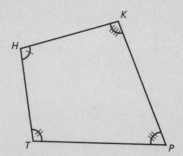

Based on the tick marks, which one of the following is correct?

(A) $\dfrac{HK}{ML} = \dfrac{KP}{IL}$ (C)   FILM ~ HTPK

(B) $\dfrac{TP}{ML} = \dfrac{HT}{FM}$ (D)   LIFM ~ KHTP

2. One square has a side of 5, and a second square has a side of 9. Which one of the following is a correct statement?

(A) The squares are congruent.

(B) The squares may be similar, if we know their perimeters.

(C) The squares are definitely similar.

(D) The squares cannot be similar.

3. Which one of the following would guarantee that quadrilateral GJMQ is similar to quadrilateral VSRN?

(A) $\angle G \cong \angle V$ and $\angle J \cong \angle S$

(B) The area of GJMQ is twice the area of VSRN.

(C) Each of the sides of GJMQ is five units larger than one of the sides of VSRN.

(D) $\angle J \cong \angle S$, $\angle M \cong \angle R$, $\angle Q \cong \angle N$, and all pairs of sides are in proportion.

 *(continued)*

4. Rectangle *AZYB* is similar to rectangle *CXDW*. If *AZ* = 30, *XD* = 6, and *DW* = 12, which one of the following is <u>completely</u> correct?

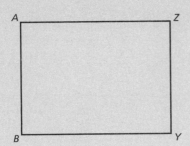

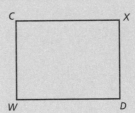

(A) *AB* = 15 and *CX* = 15   (C) *ZY* = 15 and *CX* = 12

(B) *ZY* = 30 and *CW* = 12   (D) *AB* = 30 and *CW* = 6

5. Which one of the following will guarantee that rhombus *EFGH* is <u>not</u> similar to rhombus *VUTS*?

(A) *EF* ≠ *UV*

(B) *m∠E* ≠ *m∠T*

(C) *m∠H* ≠ *m∠V*

(D) The perimeter of *EFGH* does not equal the perimeter of *VUTS*.

6. Trapezoids *IJKL* and *RQPN* are similar, as shown below.

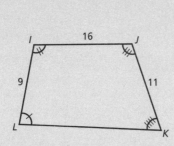

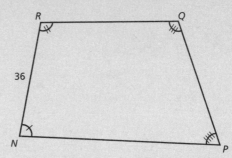

What is the value of the sum *RQ* + *QP*?

(A) 108   (C)   72

(B) 90   (D)   64

Test Yourself! (continued)

7. Isosceles trapezoid *MPRT* is similar to isosceles trapezoid *VXYZ*, and *MP* is one of the bases of the first trapezoid as shown below.

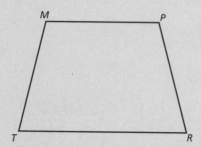
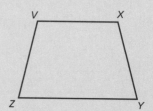

Given that *MT* = 18, *m∠R* = 80°, *TR* = 24, and *ZY* = 9.6, which one of the following is <u>completely</u> correct?

(A) *VZ* = 3.6 and *m∠Z* = 80°

(B) *XY* = 7.2 and *m∠X* = 100°

(C) *MP* = 24 and *m∠M* = 100°

(D) *VX* = 9.6 and *m∠P* = 80°

8. Similar parallelograms *ACEG* and *ZXUS* are shown below.

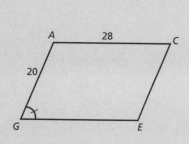
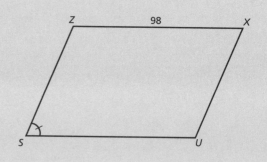

What is the value of *XU*?

Answer: _____

Perimeters of Similar Figures

In this lesson, we will explore the relationship between the perimeters of similar geometric figures. Throughout Lessons 14 and 15, we discovered the connection between the corresponding sides of similar triangles and quadrilaterals. We also determined that corresponding angles were congruent.

Your Goal: When you have completed this lesson, you should be able to determine the unknown perimeter for either of two similar figures. Also, given information concerning their perimeters, you will be able to determine the value of an unknown side. As in previous lessons, we will confine our discussion to triangles and quadrilaterals.

LESSON 16

Perimeters of Similar Figures

Consider two similar triangles, as shown below in Figures 16.1 and 16.2.

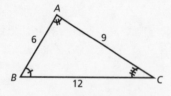

Figure 16.1

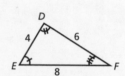

Figure 16.2

Based on only the tick marks for the angles, it is apparent that $\triangle ABC \sim \triangle DEF$. The common ratio of corresponding sides is $\frac{3}{2}$. This is easy to check since we know that $\frac{6}{4} = \frac{9}{6} = \frac{12}{8} = \frac{3}{2}$. If you were asked to take a quick guess as to the ratio of the perimeters of these triangles, your intuition would probably lean toward $\frac{3}{2}$. Your intuition would be right! Notice that the perimeter of $\triangle ABC$ is 27, and the perimeter of $\triangle DEF$ is 18. Of course, $\frac{27}{18} = \frac{3}{2}$.

The ratio of each side of $\triangle DEF$ to its corresponding side of $\triangle ABC$ is $\frac{2}{3}$.

The ratio of the perimeter of $\triangle DEF$ to that of $\triangle ABC$, represented by $\frac{18}{27}$, is also $\frac{2}{3}$.

The rule is that the ratio of the perimeters of two similar triangles equals the ratio of their corresponding sides.

Even though each of AB and DF equals 6, these are not corresponding sides. Be sure that whenever you are looking for the common ratio for any two similar figures, you check that the sides you are looking at are opposite congruent angles.

Let's consider two similar right triangles, such as Figures 16.3 and 16.4, shown below.

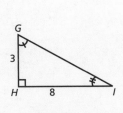

Figure 16.3

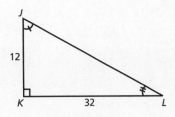

Figure 16.4

$\triangle GHI \sim \triangle JKL$, but in order to find the perimeter of each triangle, we need to determine *GI* and *JL*. We'll use the Pythagorean theorem. Let *x* represent *GI*, so that $3^2 + 8^2 = x^2$. This leads to $x^2 = 73$; thus $x = \sqrt{73} \approx 8.54$. After we add $3 + 8 + 8.54$, we see that the perimeter of $\triangle GHI \approx 19.54$. Since the common ratio between the sides of $\triangle GHI$ and $\triangle JKL$ is $\frac{3}{12} = \frac{1}{4}$, we can find the perimeter of $\triangle JKL$ by the product $(19.54)(4) = 78.16$.

Let's apply the information we learned about perimeters of triangles to perimeters of quadrilaterals. Consider two similar quadrilaterals, as shown below in Figures 16.5 and 16.6.

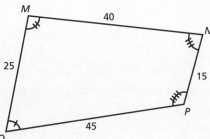

Figure 16.5

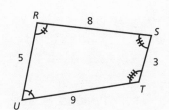

Figure 16.6

As a result of the tick marks and the numerical values assigned to the sides, *MNPQ* is similar to *RSTU*. The common ratio between corresponding sides is $\frac{5}{1} = 5$. Notice that the perimeter of *MNPQ* is 125, and the perimeter of *RSTU* is 25. Sure enough, the ratio between the perimeters is also 5.

The rule is that for any two similar quadrilaterals, the ratio of their perimeters equals the ratio of any pair of corresponding sides.

This rule will also apply to any of the "special" quadrilaterals we have studied: the square, rectangle, rhombus, parallelogram, and trapezoid. Always be careful that you match up corresponding sides.

If two figures are congruent, they are <u>always</u> considered similar. The reason is that the ratio of corresponding sides (and perimeters) is $\frac{1}{1} = 1$. However, two similar figures are usually not congruent.

1 Example: $\triangle VWX \sim \triangle YZA$. If $VW = 3$, $VX = 10$, and $YA = 35$, what is the value of YZ?

Solution: Let x represent YZ. Since $\frac{VW}{YZ} = \frac{VX}{YA}$, we can substitute values to get $\frac{3}{x} = \frac{10}{35}$. Cross-multiply to get $10x = 105$. Thus, $x = 10.5$.

2 Example: *Consider the two similar isosceles triangles shown below.*

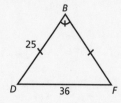 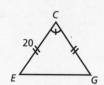

What is the perimeter of $\triangle CEG$?

Solution: First, we need to determine CG and EG.
Since $CE = CG$, we know that $CG = 20$. Let x represent EG.
Then $\frac{25}{20} = \frac{36}{x}$.
Cross-multiply to get $25x = 720$. Thus, $x = 28.8$.
Then the perimeter of $\triangle CEG$ is $20 + 20 + 28.8 = 68.8$.

As a check for the accuracy of this answer, notice that the perimeter of $\triangle BDF$ is $25 + 25 + 36 = 86$. Now we can verify that $\frac{86}{68.8} = \frac{5}{4}$, which is exactly the value of the ratio of corresponding sides.

3 Example: **HIJ and KLM are each right isosceles triangles, with right angles at I and L as shown below.**

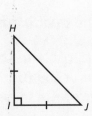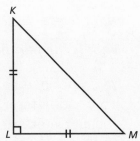

HI = 6, and the ratio of the perimeter of HIJ to that of KLM is $\frac{2}{9}$. To the nearest hundredth, what is the value of KM?

Solution: Let's first find the value of *HJ*. Letting x represent *HJ* and using the Pythagorean theorem on $\triangle HIJ$, $6^2 + 6^2 = x^2$.
This is followed by $x^2 = 72$.
Then $x = \sqrt{72} \approx 8.49$.

Since the ratio of the perimeters must equal the ratio $\frac{HJ}{KM}$, letting x represent KM, we have $\frac{2}{9} = \frac{8.49}{x}$. Cross-multiply to get $2x = 76.41$.
Thus, $x = 38.21$.

4 Example: **Consider similar right triangles NPQ and RST, as shown below.**

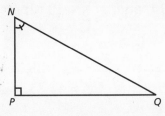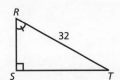

PQ is twice NP, and the ratio of the perimeter of triangle NPQ to that of triangle RST is $\frac{7}{4}$. To the nearest hundredth, what is the value of NP?

Solution:

Let's first determine *NQ*. Let x represent *NQ*. Using the given value of *RT* and the known ratio of the perimeters, we can write $\frac{7}{4} = \frac{x}{32}$.
The next step is $4x = 224$, so $x = 56$. Or, $NQ = 56$.
Focusing on $\triangle NPQ$, we know that *PQ* is twice as large as *NP*.
Let x represent *NP*, and let 2x represent *PQ*.
Using the Pythagorean theorem on $\triangle NPQ$, $x^2 + (2x)^2 = 56^2$.
The next step is $x^2 + 4x^2 = 3136$.
(Caution: Did you remember that $(2x)^2 = 4x^2$, <u>not</u> $2x^2$?)
Now we have $5x^2 = 3136$, followed by $x^2 = 627.2$.
Finally, $x = \sqrt{627.2} \approx 25.04$. Or, $NP = 25.04$.

5 **Example:**   *Two quadrilaterals are similar. The sides of the first quadrilateral are 9, 16, 30, and 35. The ratio of the perimeter of the first quadrilateral to that of the second quadrilateral is $\frac{5}{8}$. What is the perimeter of the second quadrilateral?*

Solution: We do not need a diagram for this problem. The perimeter of the first quadrilateral is simply the sum of the four given values, which is 90. Let x represent the perimeter of the second quadrilateral. Then $\frac{5}{8} = \frac{90}{x}$. Then $5x = 720$, so $x = 144$.

6 **Example:**   *Rectangles UVWX and YZAB, shown below, are similar.*

 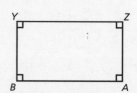

The ratio of the perimeter of UVWX to that of YZAB is 13. If WX is 39 units larger than BA, what is the value of WX?

Solution: Let x represent *BA*, and let $x + 39$ represent *WX*. Then $\frac{13}{1} = \frac{x+39}{x}$.
Cross-multiplying yields $13x = x + 39$, followed by $12x = 39$, and then $x = 3.25$. Thus, *WX* = 42.25.

7 **Example:** *Parallelogram CDEF is similar to parallelogram GHIJ, as shown below.*

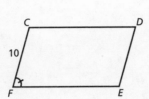 

The ratio of the perimeter of CDEF to that of GHIJ is $\frac{5}{12}$, CF = 10, and IJ is 49 units larger than EF. What is the perimeter of CDEF?

Solution: Let x represent *EF*, and let $x + 49$ represent *IJ*.

Then $\frac{x}{x+49} = \frac{5}{12}$. Cross-multiply to get $12x = 5x + 245$.

Then $7x = 245$, so $x = 35$.

The perimeter of CDEF is $(2)(10) + (2)(35) = 90$.

8 **Example:** *Trapezoid KLMN is similar to trapezoid PQRS, as shown below.*

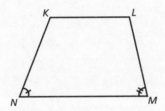 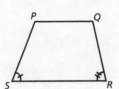

KN = 28, SP = 8, PQ = 12, and QR = 10. What is the value of KL + LM?

Solution: The ratio of the corresponding sides of *KLMN* to PQRS is $\frac{28}{8} = \frac{7}{2}$.

Let x represent *KL*. Then $\frac{x}{12} = \frac{7}{2}$, which leads to $2x = 84$.

Thus, $x = KL = 42$.

Now let x represent *LM*. Then $\frac{x}{10} = \frac{7}{2}$, so this means that $2x = 70$.

Thus, $x = LM = 35$. Finally, $KL + LM = 77$.

9   **Example:**   *Isosceles trapezoid TUVW is similar to isosceles trapezoid XYZA, as shown below.*

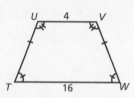
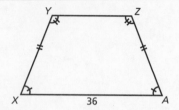

UV = 4, TW = 16, and XA = 36. If the perimeter of TUVW is 38, what is the value of XY?

Solution: Let x represent *UT*. Since *TUVW* is isosceles, x must also represent *VW*. Then $x + 4 + x + 16 = 38$.
This equation simplifies to $2x + 20 = 38$.
The next steps are $2x = 18$, followed by $x = 9$.

Since *AX* = 36, the ratio between corresponding sides of *TUVW* to *XYZA* is $\frac{16}{36} = \frac{4}{9}$.
Letting x represent *XY*, we have $\frac{9}{x} = \frac{4}{9}$. Thus, $4x = 81$.
Finally, $x = XY = 20.25$.

Test Yourself!

1.   **△BCD ~ △EFG, as shown below.**

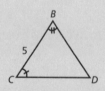
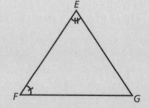

If the common ratio of corresponding sides from triangle *BCD* to triangle *EFG* is $\frac{2}{5}$, what is the value of *EF*?

Answer: _____

Test Yourself! (continued)

2. Suppose that triangle HIJ is similar to triangle *MLK*. *HI* = 11, *HJ* = 9, and *MK* = 8. Which one of the following statements <u>must</u> be true?

   (A) *IJ* is the largest side of △*HIJ*.

   (B) *IJ* > *LK*

   (C) *ML* < 8

   (D) Both triangles are scalene.

3. Right triangles *NQS* and *PRT* are similar, as shown below.

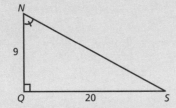
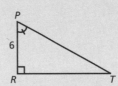

   To the nearest hundredth, what is the value of *PT*?

   Answer: _____

4. Quadrilaterals *UWYZ* and *VABD* are similar, as shown below.

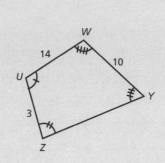
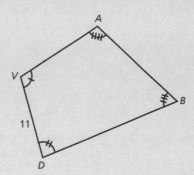

   If the perimeter of *UWYZ* is 51, what is the value of *DB*?

   Answer: _____

5. The smallest side of quadrilateral *EFGH* is 10, which is also the largest side of quadrilateral *IJKL*. Under which one of the following circumstances could these quadrilaterals be similar?

    (A) They cannot be similar.

    (B) The largest side of *EFGH* is 20, and the smallest side of *IJKL* is 4.

    (C) Two corresponding sides of these quadrilaterals are equal.

    (D) The smallest side of *IJKL* is 4, and the largest side of *EFGH* is 25.

6. Right triangles *MNP* and *QRS* are similar, as shown below.

    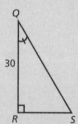

    MP is three times as large as *NP*, and the ratio of the perimeter of $\triangle MNP$ to $\triangle QRS$ is $\frac{5}{6}$. To the nearest hundredth, what is the value of *NP*?

    Answer: _____

7. Two quadrilaterals are similar. The sides of the first quadrilateral are 17, 22, 31, and 40. The largest side of the second quadrilateral is 100. What is the perimeter of the second quadrilateral?

    Answer: _____

Test Yourself! (continued)

8. Rectangles *TVAB* and *UXCD*, shown below, are similar.

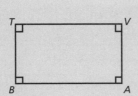

The ratio of the perimeter of *TVAB* to *UXCD* is $\frac{7}{10}$. If *CD* is 9 units larger than *AB*, what is the value of *CD*?

Answer:_____

9. Which two of the following are <u>always</u> similar?

(A) Two rectangles   (D) Two isosceles triangles

(B) Two squares   (E) Two equilateral triangles

(C) Two right triangles   (F) Two isosceles trapezoids

Answer:_____

10. Parallelograms *EGJL* and *FHIM* are similar, as shown below.

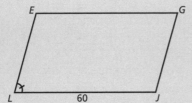

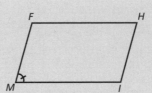

The ratio of the perimeter of *EGJL* to *FHIM* is $\frac{15}{8}$. Also, *EL* is 21 units larger than *FM*. What is the perimeter of *FHIM*?

Answer:_____

Areas of Similar Triangles

In this lesson, we will explore the relationship between the areas of similar triangles. In Lesson 16, we discovered the connection between the corresponding perimeters of similar triangles (and quadrilaterals).

Your Goal: When you have completed this lesson, you should be able to determine the unknown area for either of two similar triangles. Also, given information concerning their areas, you will be able to determine the value of an unknown side.

LESSON 17

Areas of Similar Triangles

Let's Review

SEE LESSON 8

The **area of a triangle** is given by the formula $A = \frac{1}{2}bh$, where A = area, b = base (which can be any of the three sides), and h = the corresponding height. We already know that the ratio of the perimeters of two similar triangles equals the ratio of corresponding sides.

Now, let's discuss the relationship between two corresponding altitudes of similar triangles. Consider Figures 17.1 and 17.2, which are similar triangles that are intentionally drawn to scale.

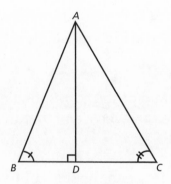

Figure 17.1

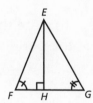

Figure 17.2

$\triangle ABC \sim \triangle EFG$, with the ratio of the corresponding sides equal to $\frac{2}{1}$. Use your ruler to check the lengths of altitudes $\overline{AD}$ and $\overline{EH}$. Hopefully, you will find that $\overline{AD}$ is twice as large as $\overline{EH}$. It certainly seems reasonable that the ratio of corresponding heights is the same as the ratio of the corresponding sides.

MathFlash!

When we refer to the sides of a triangle, the word "side" can refer to either the line segment or to the length of this line segment. The word "altitude" is normally used for the actual line segment, whereas "height" is used for its actual length.

The triangles used in Figures 17.1 and 17.2 are acute. Let's use another illustration in which both similar triangles are obtuse, as seen in Figures 17.3 and 17.4, shown below.

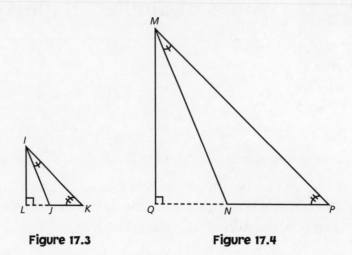

Figure 17.3   Figure 17.4

$\triangle IJK \sim \triangle MNP$, with the ratio of corresponding sides equal to $\frac{1}{3}$. Use your ruler to check the lengths of altitudes $\overline{IL}$ and $\overline{MQ}$. You should find that $MQ = (3)(IL)$.

> Once again, it appears that the ratio of the corresponding heights is the same as the ratio of the corresponding sides.
>
> We can extend this concept to state that the ratio of the corresponding heights is also equal to the ratio of the perimeters. We can even claim that this is true for quadrilaterals.

We are now ready to investigate the relationship between the sides of two similar triangles and their associated areas.

Suppose $\triangle RST$ is similar to $\triangle UVW$, as shown below in Figures 17.5 and 17.6.

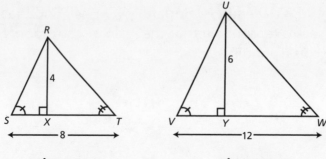

Figure 17.5   Figure 17.6   (Figures not drawn to scale)

Notice that the ratio of corresponding sides from $\triangle RST$ to $\triangle UVW$ is $\frac{4}{6}=\frac{8}{12}=\frac{2}{3}$.

The area of $\triangle RST$ is $\left(\frac{1}{2}\right)(4)(8)=16$, and the area of $\triangle UVW$ is $\left(\frac{1}{2}\right)(6)(12)=36$.

So the ratio of areas is $\frac{16}{36}=\frac{4}{9}$. Of course $\frac{4}{9}\neq\frac{2}{3}$, but they are related!

Hopefully, you can see that $\left(\frac{2}{3}\right)^2=\frac{4}{9}$.

Let's try a second problem, without a diagram. Triangle 1 has a base of 15 and a height of 7. Triangle 2 is similar to Triangle 1. Its corresponding base is 60, and its corresponding height is 28.

The ratio of the corresponding sides from Triangle 1 to Triangle 2 is $\frac{15}{60}=\frac{7}{28}=\frac{1}{4}$.

Now the area of Triangle 1 is $\left(\frac{1}{2}\right)(15)(7)=52.5$,

and the area of Triangle 2 is $\left(\frac{1}{2}\right)(60)(28)=840$.

What is the reduced fraction for $\frac{52.5}{840}$?

First rewrite 52.5 as $\frac{105}{2}$.

Then we have $\frac{105}{2}\div840=\left(\frac{105}{2}\right)\left(\frac{1}{840}\right)$.

$\left(\frac{\overset{1}{105}}{2}\right)\left(\frac{1}{\underset{8}{840}}\right)=\frac{1}{16}$. Comparing the numbers $\frac{1}{4}$ and $\frac{1}{16}$, we have $\left(\frac{1}{4}\right)^2=\frac{1}{16}$.

In conclusion:

- The **ratio of the areas** is the square of the ratio of the corresponding *sides*.
- The **ratio of the areas** is the square of the ratio of the corresponding *heights*.
- The **ratio of the areas** is the square of the ratio of the *perimeters*.

1   **Example:**   *Triangle ABC is similar to triangle DEF, as shown below.*

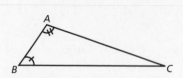 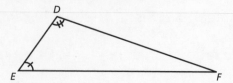

The perimeters of ABC and DEF are 10 and 25, respectively. If the area of △ABC is 30, what is the area of △DEF?

Solution: Let x represent the area of △DEF. Then, $\left(\dfrac{10}{25}\right)^2 = \dfrac{30}{x}$, which becomes $\dfrac{100}{625} = \dfrac{30}{x}$.

Cross-multiply to get $100x = 18{,}750$. Thus, $x = 187.5$.

MathFlash!

It would certainly be permissible to reduce $\dfrac{10}{25}$ to $\dfrac{2}{5}$ before squaring. Then the equation would be $\left(\dfrac{2}{5}\right)^2 = \dfrac{4}{25} = \dfrac{30}{x}$. The equation $\dfrac{4}{25} = \dfrac{30}{x}$ is easier to solve.

2   **Example:**   *Triangle GHI is similar to triangle JKL, as shown below. The area of △GHI is 60, and the area of △JKL is 40. If GM = 8, then what is the value, to the nearest hundredth, of JN?*

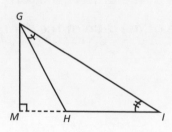 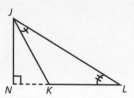

Solution: Let x represent JN. The area ratio is $\dfrac{60}{40}$, which reduces to $\dfrac{3}{2}$.

Now, since the ratio of the areas equals the square of the ratio of the altitudes, we can write $\left(\dfrac{8}{x}\right)^2 = \dfrac{3}{2}$.

This equation simplifies to $\dfrac{64}{x^2} = \dfrac{3}{2}$.

Cross-multiply to get $3x^2 = 128$.

Dividing by 3 yields $x^2 \approx 42.67$.

Finally, $x = \sqrt{42.67} \approx 6.53$.

3 **Example:** *Triangle MNP is similar to triangle QRS. The area of △MNP is 30 and the area of △QRS is 42. If QS = 14, what is the value, to the nearest hundredth, of MP?*

Solution: Let's try to solve this problem without a diagram. Let x represent MP. Our ratio formula will be $\left(\dfrac{MP}{QS}\right)^2 = \dfrac{30}{42}$. Substitute and reduce $\dfrac{30}{42}$ to $\dfrac{5}{7}$ to get $\left(\dfrac{x}{14}\right)^2 = \dfrac{5}{7}$. Squaring the left side yields $\dfrac{x^2}{196} = \dfrac{5}{7}$. Cross-multiply to get $7x^2 = 980$. Now divide both sides by 7 to get $x^2 = 140$. Finally, $x = \sqrt{140} \approx 11.83$.

Thus far, the only way we can determine the area of a triangle is to use the formula $A = \left(\dfrac{1}{2}\right)(b)(h)$. This means that knowing the value of one side of a triangle, but not knowing the value of its corresponding height, would seem insufficient to determining the area.

Luckily, an individual named Heron discovered a magic formula that can determine the area of any triangle, provided that we know the lengths of all three sides. Here is **Heron's formula**: $A = \sqrt{s(s-a)(s-b)(s-c)}$. In this formula, A is the area; a, b, and c represent the sides (in no particular order of size); and s represents half the perimeter. In most cases when you use this formula, you will round off your answer to the nearest hundredth.

4 **Example:** *What is the area of a triangle whose sides are 5, 6, and 7?*

Solution: $a = 5$, $b = 6$, and $c = 7$. The value of s can be found by evaluating $\left(\dfrac{1}{2}\right)(5+6+7) = \left(\dfrac{1}{2}\right)(18) = 9$. Now simply substitute the known values into the formula to get $A = \sqrt{9(9-5)(9-6)(9-7)} = \sqrt{9(4)(3)(2)} = \sqrt{216} \approx 14.70$.

5 **Example:** *What is the area of a triangle whose sides are 12, 21, and 16?*

Solution: $a = 12$, $b = 21$, and $c = 16$. (Remember that a, b, and c are interchangeable in Heron's formula.)

Also, $s = \left(\dfrac{1}{2}\right)(12 + 21 + 16) = 24.5$.

Thus, the area of the triangle is

$$\sqrt{(24.5)(24.5 - 12)(24.5 - 21)(21.5 - 16)} =$$
$$\sqrt{(24.5)(12.5)(3.5)(5.5)} \approx \sqrt{5895.31} \approx 76.78.$$

6 **Example:** *Let's apply Heron's formula to a problem involving similar triangles. Suppose △TUV is similar to △XYZ, as shown below.*

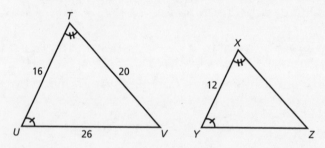

What is the area of △XYZ?

Solution: Using Heron's formula for $\triangle TUV$, $a = 16$, $b = 20$, $c = 26$, and $s = \left(\dfrac{1}{2}\right)(16 + 20 + 26) = 31$. Then its area is $\sqrt{(31)(15)(11)(5)} = \sqrt{25{,}575} \approx 159.92$. Let x represent the area of $\triangle XYZ$. We know that the ratio of the areas equals the square of the ratio of corresponding sides. If we reduce $\dfrac{16}{12}$ to $\dfrac{4}{3}$, we can write $\left(\dfrac{4}{3}\right)^2 = \dfrac{159.92}{x}$. The equation then becomes $\dfrac{16}{9} = \dfrac{159.92}{x}$.

Cross-multiply to get $16x = 1439.28$. Finally, $x \approx 89.96$.

The use of Heron's formula will automatically work on "special" triangles, such as those that are isosceles, equilateral, right, and so forth. Let's show how this formula works for two similar right triangles.

7 **Example:**   *Right triangles ABC and DEF are similar, as shown below.*

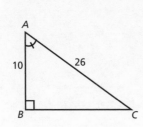
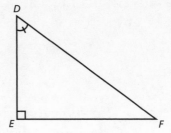

If the area of △DEF is 270, what is the value of EF?

Solution: This problem looks impossible, but it is not! Let's first find the area of △ABC. Letting x represent BC, use the Pythagorean theorem. Then $10^2 + x^2 = 26^2$. This leads to $100 + x^2 = 676$, followed by $x^2 = 576$, then $x = \sqrt{576} = 24$.

Now, we can use Heron's formula to find the area of △ABC. Since its half perimeter is $\left(\dfrac{1}{2}\right)(10 + 24 + 26) = 30$, the area will be $x = \sqrt{(30)(20)(6)(4)} = \sqrt{14,400} = 120$.

Now let x represent EF. Since the ratio of the areas is equal to the square of the ratio of corresponding sides, we write $\left(\dfrac{24}{x}\right)^2 = \dfrac{120}{270}$. Reduce the fraction $\dfrac{120}{270}$ to $\dfrac{4}{9}$. The equation then becomes $\dfrac{576}{x^2} = \dfrac{4}{9}$.

Cross-multiply to get $4x^2 = 5184$, followed by $x^2 = 1296$, and finally $x = 36$.

MathFlash!

Example 7 was unusually hard. However, can you see a way to solve it without using Heron's formula? There actually is a way this can be done! After we found that $BC = 24$, we can find the area of $\triangle ABC$ by using Area $= \left(\dfrac{1}{2}\right)(24)(10) = 120$. This is only allowed because there is a right angle at B. After calculating the area of $\triangle ABC$, the rest of the solution would be identical to the solution shown above.

8 **Example:**   *Isosceles triangles GHI and JKL are similar, as shown below.*

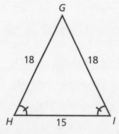

What is the area of △JKL?

Solution: Let's make this solution as painless as possible. We already know that $JL = 6$. Let x represent KL, and write $\dfrac{18}{6} = \dfrac{15}{x}$.
Then $18x = 90$, so $x = 5$.

All that is left is to apply Heron's formula to $\triangle JKL$, in which the half perimeter is $\left(\dfrac{1}{2}\right)(6+6+5) = 8.5$.

Letting $a = b = 6$, and $c = 5$, the area of $\triangle JKL$ becomes

$$\sqrt{(8.5)(2.5)(2.5)(3.5)} \approx \sqrt{185.94} \approx 13.64.$$

Test Yourself!

1. It is known that the area of Triangle 1 is 9, and the area of Triangle 2 is 25. Which one of the following could represent the lengths of corresponding sides?

   (A)   18 and 50

   (B)   12 and 28

   (C)   6 and 10

   (D)   3 and 6

2. Triangles *ACE* and *GIJ* are similar. The perimeter of △*ACE* is 40, and the perimeter of △*GIJ* is 75. If the area of △*ACE* is 60, what is the area of △*GIJ*?

   Answer: _____

3. Triangles *KMP* and *BDF* are similar, as shown below.

   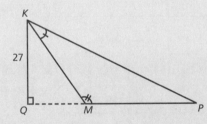
   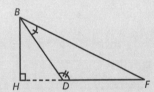

   The ratio of the area of △*KMP* to that of △*BDF* is $\frac{9}{5}$. To the nearest hundredth, what is the value of *BH*?

   Answer: _____

4. The shortest side of the first of two similar triangles is 4. The second triangle has sides of 20, 14, and 8. What is the ratio of the area from the first triangle to the area of the second triangle?

   (A) $\frac{1}{2}$

   (B) $\frac{1}{4}$

   (C) $\frac{1}{5}$

   (D) $\frac{1}{25}$

Test Yourself! (continued)

5.   To the nearest hundredth, what is the area of a triangle whose sides are 8, 15, and 21?

   Answer: _____

6.   Isosceles triangle *QRS* is shown below.

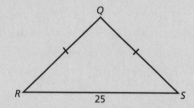

   The perimeter of △*QRS* is 57. To the nearest hundredth, what is the area of this triangle?

   Answer: _____

7.   Triangles *TUV* and *WXY* are both equilateral. *TU* = 12, and the perimeter of △*WXY* is 30. Which one of the following represents the ratio of the area from △*TUV* to △*WXY*?

   (A) $\dfrac{4}{25}$ (C) $\dfrac{6}{5}$

   (B) $\dfrac{2}{5}$ (D) $\dfrac{36}{25}$

8.   Triangles *BFG* and *CHJ* are similar, as shown below.

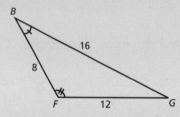 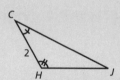

   To the nearest hundredth, what is the area of △*CHJ*?

   Answer: _____

Test Yourself! (continued)

9.   Given two similar triangles, suppose that the area of the smaller triangle is 20 and the area of the larger triangle is 70. If the largest side of the smaller triangle is 18, then what is the largest side of the larger triangle to the nearest hundredth?

Answer: _____

10.   Right triangles *KLM* and *NPQ* are similar, as shown below.

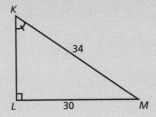
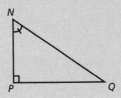

If the area of △*NPQ* is 200, what is the value of *NP* to the nearest hundredth?

Answer: _____

Areas of Similar Quadrilaterals

In this lesson, we will explore the relationship between the areas of similar quadrilaterals. In Lesson 17, we discovered the connection between the corresponding areas of similar triangles. Be sure you completely understand the concepts of Lesson 17, as you will need them in this lesson.

Your Goal: When you have completed this lesson, you should be able to determine the unknown area for either of two similar quadrilaterals. Also, given information concerning their areas, you will be able to determine the value of an unknown side.

Areas of Similar Quadrilaterals

The ratio of the areas of two similar quadrilaterals is equal to the square of the ratio of both their corresponding sides and corresponding perimeters. This is the same conclusion that we reached in Lesson 17 concerning similar triangles.

1 **Example:** *Consider two similar quadrilaterals, ABCD and EFGH, as shown below in Figures 18.1 and 18.2.*

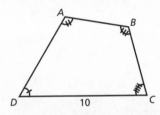

Figure 18.1

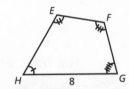

Figure 18.2

If the area of EFGH is 100, what is the area of ABCD?

Solution: The ratio of corresponding sides is $\frac{10}{8}$, which reduces to $\frac{5}{4}$.

Let x represent the area of *ABCD*. Then $\left(\frac{5}{4}\right)^2 = \frac{x}{100}$ which becomes

$\frac{25}{16} = \frac{x}{100}$. Next $16x = 2500$, so that $x = 156.25$.

2 **Example:** *Quadrilateral IJKL is similar to quadrilateral MNPQ. The area of IJKL is 180, and the area of MNPQ is 400. If KL = 13, what is the value of PQ to the nearest hundredth?*

Solution: The ratio of the areas is $\frac{180}{400}$. Reduce this to $\frac{9}{20}$. Let x represent *PQ*.

Then $\left(\frac{13}{x}\right)^2 = \frac{9}{20}$, followed by $\frac{169}{x^2} = \frac{9}{20}$. Next is $9x^2 = 3380$. Divide

by 9 to get $x^2 \approx 375.56$. Finally, $x = \sqrt{375.56} \approx 19.38$.

Let's now discuss the **special quadrilaterals**. We will begin with two **squares**, which are always similar. If the side of Square 1 is 6, and the side of Square 2 is 12, we know that the ratio of their sides, $\frac{6}{12}$, reduces to $\frac{1}{2}$. The ratio of their perimeters, which are 24 and 48, also reduces to $\frac{1}{2}$.

We can also easily verify that the ratio of their areas is $\left(\frac{1}{2}\right)^2 = \frac{1}{4}$. Notice that the area of Square 1 is 36, and the area of Square 2 is 144. Sure enough, $\frac{36}{144}$ reduces to $\frac{1}{4}$.

3 **Example:** *Consider Square #3 and Square #4, as shown below.*

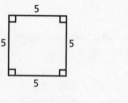

Square 3   Square 4

The ratio of the area of Square 3 to the area of Square 4 is $\frac{3}{5}$.
What is the perimeter of Square 4?

Solution: The perimeter of Square 3 is $(5)(4) = 20$. Let x represent the perimeter of Square 4. Then $\left(\frac{20}{x}\right)^2 = \frac{3}{5}$, followed by $\frac{400}{x^2} = \frac{3}{5}$. Cross-multiply to get $3x^2 = 2000$. This leads to $x^2 \approx 666.67$, which means that $x = \sqrt{666.67} \approx 25.82$.

MathFlash!

In order to check this answer, we can determine that each side of Square 4 is approximately equal to $\frac{25.82}{4} = 6.455$. This means that the area of Square 4 is approximately $(6.455)^2 \approx 41.67$. Finally, notice that the ratio $\frac{25}{41.67}$ is __extremely__ close to $\frac{3}{5}$.

Now consider two similar **rectangles**, as shown below, in Figures 18.3 and 18.4.

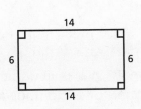

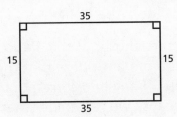

Figure 18.3   Figure 18.4

We can easily show that they are similar, since $\frac{6}{15} = \frac{14}{35} = \frac{2}{5}$. Their respective areas are $(14)(6) = 84$ and $(35)(15) = 525$. Note that the ratio of their areas, $\frac{84}{525}$, reduces to $\frac{4}{25}$. Sure enough, $\left(\frac{2}{5}\right)^2 = \frac{4}{25}$. We see again that the ratio of the areas is the square of the ratio of the sides.

4 **Example:**   *Consider two similar rectangles, as shown below.*

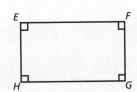

If the ratio of the areas from ABCD to EFGH is $\frac{16}{9}$ and AD is 22 units larger than EH, what is the value of AD?

Solution: Let x represent *EH*, so that $x + 22$ represents *AD*.

Since the ratio of areas is $\frac{16}{9}$, the ratio of sides must be $\sqrt{\frac{16}{9}} = \frac{4}{3}$.

Now, we can write $\frac{x+22}{x} = \frac{4}{3}$.

Next $4x = 3x + 66$, followed by $x = 66$. Thus, *AD = 88*.

We have used the fact that the ratio of the sides must equal the square root of the ratio of the areas.

For a **rhombus**, the length of a side is usually given, but its height is normally not provided. To find the area of a rhombus we use the formula $A = \left(\dfrac{1}{2}\right)(d_1)(d_2)$, in which d_1 and d_2 represent the lengths of the diagonals. Since diagonals are in the same "league" as sides, you can anticipate that the ratio of the areas of any two rhombi is equal to the square of any two <u>corresponding</u> diagonals. We have underscored "corresponding" because we remember that the diagonals of a rhombus are not equal.

5 **Example:** *Consider similar rhombi IJKL and MNPQ, as shown below.*

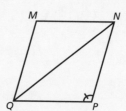

If the ratio of the areas from IJKL to MNPQ is $\dfrac{4}{7}$, and QN is 10 units larger than JL, what is the value of JL?

Solution: Let x represent *JL*, so that $x + 10$ represents *QN*. The ratio of the diagonals is equal to $\sqrt{\dfrac{4}{7}} \approx 0.756$.

Now, we can write $\dfrac{x}{x+10} = \dfrac{0.756}{1}$.

Cross-multiply to get $x = 0.756x + 7.56$. Subtract $0.756x$ from each side of the equation to get $0.244x = 7.56$. Finally, $x \approx 30.98$.

MathFlash!

Let's check our answer. We now know that QN $\approx 30.98 + 10 \approx 40.98$.

So, the ratio of the corresponding diagonals is $\dfrac{30.98}{40.98}$. This implies that the ratio of the areas should be $\left(\dfrac{30.98}{40.98}\right)^2$, which is very close to $\dfrac{4}{7}$.

6 **Example:** *Refer back to Example 5. Suppose IK = 8. What is the area of rhombus MNPQ?*

Solution: First find the value of *MP*. Let *x* represent *MP*. The ratio of the corresponding diagonals ≈ 0.756, so we can write $\dfrac{8}{x} = \dfrac{0.756}{1}$. Cross-multipy to get $0.756x = 8$. So $x \approx 10.58$. Now, the area of *MNPQ* is $\left(\dfrac{1}{2}\right)(MP)(QN) = \left(\dfrac{1}{2}\right)(10.58)(40.98) \approx 216.78$.

MathFlash!

You can verify that the area of rhombus *IJKL* $\approx \left(\dfrac{1}{2}\right)(8)(30.98) = 123.92$. Furthermore, notice that $\dfrac{123.92}{216.78}$ is, again, <u>very</u> close to the desired fraction $\dfrac{4}{7}$.

Now let's turn our attention to **trapezoids**.

7 **Example:** *Consider two similar trapezoids RSTU and WXYZ, as shown below.*

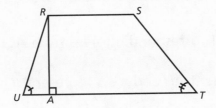 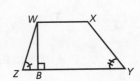

If RA = 30, WB = 21, and the area of RSTU is 170 square units larger than WXYZ, what is the area of trapezoid WXYZ?

Solution: The ratio of corresponding sides equals that of corresponding heights, which is $\dfrac{30}{21} = \dfrac{10}{7}$. Let *x* represent the area of *WXYZ* and $x + 170$ represent the area of *RSTU*. Then, our proportion becomes $\left(\dfrac{10}{7}\right)^2 = \dfrac{x+170}{x}$. Rewrite the left side as $\dfrac{100}{49}$. Then $\dfrac{100}{49} = \dfrac{x+170}{x}$.

Cross-multiply to get $100x = 49x + 8330$. Then $51x = 8330$. Thus, $x \approx 163.33$.

Example: *Refer back to Example 7. Suppose RS = 12. What is the value of WX?*

Solution: We will represent *WX* as *x*. The ratio of the heights must equal the ratio of any corresponding sides, so $\dfrac{30}{21} = \dfrac{12}{x}$.

Then $30x = 252$, so $x = 8.4$. So the value of *WX* is 8.4.

The **ratio of the corresponding sides** is the <u>square root</u> of the ratio of the <u>areas</u>.

 Example: *Isosceles trapezoids ACEG and JLNP are shown below.*

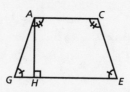 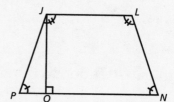

The area of *ACEG* is 64, and the area of *JLNP* is 110. If *JQ* is 5 units larger than *AH*, what is the value of *JQ*?

Solution: The ratio of corresponding sides from *ACEG* to *JLNP* is $\sqrt{\dfrac{64}{110}} \approx 0.76$.

Let *x* represent *AH* and *x* + 5 represent *JQ*. Now write $\dfrac{x}{x+5} = \dfrac{0.76}{1}$.

Then $x = 0.76x + 3.8$, followed by $0.24x = 3.8$.
So, $x \approx 15.83$. Finally, *JQ* is $15.83 + 5 = 20.83$.

MathFlash!

In Example 9, many numbers were rounded off. If you waited to round off until the last step, your final answer would be closer to 8.05. Even a difference of 0.03 for this type of problem is not a real concern. Also, don't worry if the height is much larger than the bases of a trapezoid.

1. For which one of the following is it <u>impossible</u> for Parallelogram 1 to be similar to Parallelogram 2?

    (A) Parallelogram 1: length of 18 and area of 81
         Parallelogram 2: length of 12 and area of 36

    (B) Parallelogram 1: length of 15 and area of 90
         Parallelogram 2: length of 25 and area of 250

    (C) Parallelogram 1: length of 24 and area of 256
         Parallelogram 2: length of 21 and area of 196

    (D) Parallelogram 1: length of 44 and area of 200
         Parallelogram 2: length of 99 and area of 300

2. Quadrilateral *MNPQ* is similar to quadrilateral *RSTU*. *MN* = 16, *RS* = 56, and the area of *RSTU* is 300. What is the area of *MNPQ* to the nearest hundredth?

    Answer: _____

3. Each side of Square 1 is 9 units. The ratio of the area of Square 1 to Square 2 is $\frac{20}{13}$, what is the perimeter, in units, of Square 2?

    Answer: _____

4. Consider the following two similar rectangles, as shown below.

 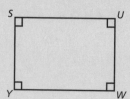

SY is 14 units larger than RX, and the ratio of the areas of RTVX to SUWY is $\frac{25}{49}$. What is the value of RX?

Answer: _____

5. Consider the following two similar rhombi, as shown below.

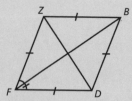 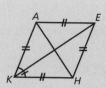

The area of ZBDF is 840 square units, and the area of AEHK is 150 square units. If EK = 18 units, what is the value, in units, of FB?

Answer: _____

6. Using the information from question 5, including the value of FB, what is the value, in units, of ZD?

Answer: _____

7. Rectangles LMNP and QRST are similar. LM = 59, and the perimeter of LMNP is 189. The ratio of the areas of LMNP to QRST is $\frac{10}{1}$. What is the value of QT to the nearest hundredth?

Answer: _____

Test Yourself! (continued)

8. Besides two squares, which of the following are always similar figures?

   (A)   Two rectangles, each of which has a length of 20

   (B)   Two rhombi, each of which has an angle measure of 80°

   (C)   Two parallelograms, each of which has a length of 26 and a width of 13

   (D)   Two isosceles trapezoids

9. Consider the following two similar trapezoids, as shown below.

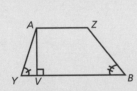 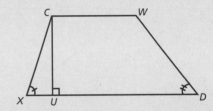

   The area of *AZBY* is 144 square units, and the area of *CWDX* is 200 square units. If *AV* = 10 units, what is the sum, to the nearest unit, of *CW* and *XD*?

   Answer: _____

10. Two similar isosceles trapezoids *EFGH* and *QRST* are shown below.

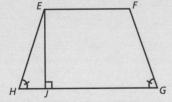 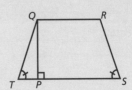

   EF = 18, *EJ* = 6, *GH* = 21.5, and the area of *EFGH* is 28 square units larger than the area of *QRST*. To the nearest hundredth, what is the value of *QR*?

   Answer: _____

QUIZ FOUR

1. The areas of similar triangles are 100 and 169, respectively. If the perimeter of the smaller triangle is 15, what is the perimeter of the larger triangle?

   A 17.25

   B 19.5

   C 25.35

   D 42

2. Quadrilateral *ABCD* is similar to quadrilateral *GHJK*, as shown below.

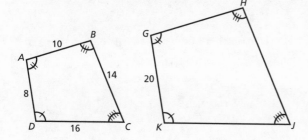

   How much larger is *JK* than *GH*?

   A 40

   B 30

   C 15

   D 5

3. Rectangle *LMNP* is <u>not</u> similar to rectangle *QRST*. *LM* = 18 and *MN* = 8. Which one of the following could represent the values of a pair of sides in rectangle *QRST*?

   A *QR* = 26 and *RS* = 12

   B *ST* = 36 and *QT* = 16

   C *QR* = 4.5 and *QT* = 2

   D *ST* = 63 and *RS* = 28

4. Right triangle *UVW* is <u>similar</u> to right triangle *XYZ*, as shown below.

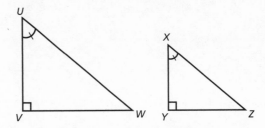

   If *UV* = 18, *VW* = 27, and *XY* = 12, what is the value of *XZ* to the nearest hundredth?

   A 13.42

   B 17.54

   C 21.63

   D 32.45

5. Parallelogram *ADGJ* is similar to parallelogram *CFHL*, as shown below.

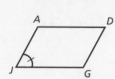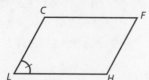

The ratio of the perimeter of *ADGJ* to that of *CFHL* is $\frac{4}{9}$. *AJ* = 8, and *HL* is 25 units larger than *GJ*. What is the perimeter of *CFHL*?

A 126

B 136

C 146

D 156

6. Consider the following two triangles.

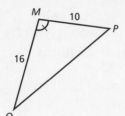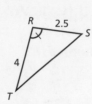

△*MPQ* ~ △*RST* by which rule?

A Side-Angle-Side

B Side-Side-Angle

C Angle-Angle-Angle

D Angle-Side-Angle

7. To the nearest hundredth, what is the area of a triangle whose sides are 18, 11, and 15?

A 35.10

B 50.84

C 66.58

D 82.32

8. Quadrilateral *UVWX* is similar to quadrilateral *YZAB*. The area of *UVWX* is 250 square inches, and the area of *YZAB* is 175 square inches. If *YZ* = 9, then what is the value of *UV* to the nearest hundredth?

A 9.71 inches

B 10.76 inches

C 11.81 inches

D 12.86 inches

9. Which one of the following would guarantee that rhombus *CDEF* is similar to but not congruent to rhombus *GHJK*?

A *EF* = 30, *JK* = 60, *m∠D* = 100°, and *m∠J* = 100°

B *DE* = 25, *HJ* = 25, *m∠C* = 35°, and *m∠G* = 55°

C *CF* = 40, *GK* = 40, *m∠E* = 75°, and *m∠J* = 75°

D *CD* = 18, *GH* = 63, *m∠F* = 110°, and *m∠G* = 70°

10. Trapezoid *LMNP* is similar to trapezoid *QRST*, as shown below.

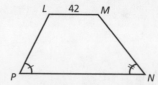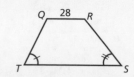

If the area of *LMNP* is 900 square units more than the area of *QRST*, what is the area of trapezoid *LMNP*?

A 720

B 1200

C 1620

D 1800

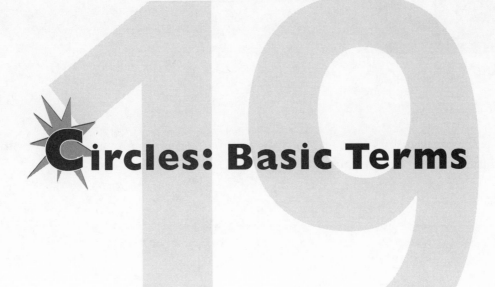

Circles: Basic Terms

In this lesson, we will explore the basic terms associated with a circle. You have certainly seen many physical objects that are circular, such as (a) automobile tires, (b) oranges, (c) baseballs, and (d) coffee cup lids.

Your Goal: When you have completed this lesson, you should be able to define the terms that are related to a circle, as well as understand how these terms relate to each other.

Circles: Basic Terms

A **circle** is a set of all points that are located at a fixed distance from a given point. Consider Figure 19.1, shown below.

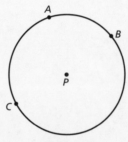

Figure 19.1

Point *P* is considered the center of this circle, although it does not lie on the circle. Points *A*, *B*, and *C* lie on the circle. Traditionally, a circle is named by its center. Thus, Figure 19.1 illustrates circle *P*.

A **radius** is a line segment whose end points are a point on the circle and the center. Consider Figure 19.2, shown below.

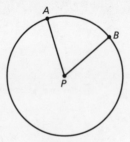

Figure 19.2

Each of $\overline{AP}$ and $\overline{BP}$ represent a radius. The plural of "radius" is "radii." (Incidentally, "radii" is one of the few words in the English language that ends in double "i's.")

A **diameter** is a line segment that passes through the center of the circle and whose end points are points on the circle. Consider Figure 19.3, shown below.

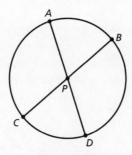

Figure 19.3

Each of $\overline{AD}$ and $\overline{BC}$ represents a diameter.

An **arc** is a part of the circle. It consists of an unbroken curve between two given points on the circle. If the endpoints of an arc correspond to the endpoints of a diameter, then the arc is called a "semicircle." If an arc is smaller than a semicircle, it is called a **minor arc**. If an arc is larger than a semicircle, it is called a **major arc**. Consider Figure 19.4, shown below.

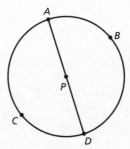

Figure 19.4

$\overset{\frown}{AB}$ represents the minor arc that extends from *A* to *B*.
$\overset{\frown}{ABD}$ represents the semicircle that contains point *B*, with endpoints of *A* and *D*.
$\overset{\frown}{ABC}$ represents the major arc that contains point *B* and extends from *A* to *C*.

MathFlash!

Minor arcs are named using just two points. For absolute clarity, semicircles and major arcs are named using three points. In naming any arc, the letters used are reversible but the middle letter used must not be an endpoint. Thus, $\overset{\frown}{AB} = \overset{\frown}{BA}$, $\overset{\frown}{ABC} = \overset{\frown}{CBA}$, and $\overset{\frown}{ABC} = \overset{\frown}{ADC}$.

A **chord** is a line segment whose endpoints lie on the circle. A diameter is really a special type of chord. Consider Figure 19.5, shown below.

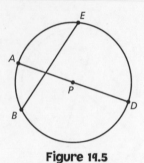

Figure 19.5

Each of $\overline{AD}$ and $\overline{BE}$ represents a chord.

A **tangent** is a line that intersects the circle exactly once.
A **secant** is a line that contains a chord of a circle. By definition, a secant intersects the circle exactly twice.

Consider Figure 19.6, shown below.

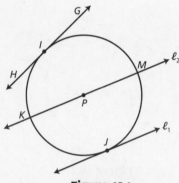

Figure 19.6

Each of $\overleftrightarrow{GH}$ and line ℓ_1 represents a tangent to circle P. The intersection point of the tangent and the circle is called the **point of tangency**. Thus, points I and J are the points of tangency for $\overleftrightarrow{GH}$ and line ℓ_1, respectively.
Line ℓ_2 represents a secant, and it contains the chord $\overline{KM}$.

A **central angle** is an angle formed by two radii of a circle.
An **inscribed angle** is an angle formed by two chords that intersect at a point on the circle. Note that one of the chords may be a diameter.

Consider Figure 19.7, shown below.

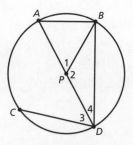

Figure 19.7

Both ∠1 and ∠2 are central angles. They intercept arcs $\overset{\frown}{AB}$ and $\overset{\frown}{BD}$, respectively. ∠3, ∠4, and ∠ABD are inscribed angles. They intercept arcs $\overset{\frown}{AC}$, $\overset{\frown}{AB}$, and $\overset{\frown}{ACD}$, respectively.

MathFlash!

The terms "radius," "diameter," "chord," and "arc" can refer to either physical appearance or to length. The interpretation depends on the context in which a term is used.

We already know that angles are measured in degrees and that line segments, such as radii, diameters, and chords, are measured in linear units (inches, centimeters, etc.).

Arcs may be measured in one of two ways, namely, in linear units or in degrees. In order to measure an arc in linear units, just imagine that the arc is "stretched out" as if it were a line segment. **The measure of an arc in degrees is defined as the measure of the central angle that intercepts it**. Look at Figure 19.8, shown below.

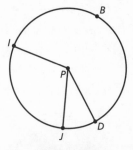

Figure 19.8

The **circumference** of a circle refers to the circle's entire length, and its measure is 360°.

A **reflex angle** is an angle whose measure is greater than 180° but less than 360°. We know that angles can only be named with three letters at most, but there is a way to avoid confusion when dealing with reflex angles. As an example, whenever we refer to ∠*IPD* in Figure 19.8, we mean the angle that is less than 180°. This means that corresponding arc is $\overarc{IJD}$. But if we are referring to reflex ∠*IPD* in Figure 19.8, then we mean the angle that is greater than 180°. Thus, for reflex ∠*IPD*, the corresponding arc is $\overarc{IBD}$.

MathFlash!

You will only be working with a reflex angle if the wording of the example specifically uses the term "reflex." Also, notice that any arc associated with a reflex angle is larger than a semicircle, which means larger than 180°. Remember that a full circle has a measure of 360°.

1 **Example:** *Refer back to Figure 19.8. If the measure of reflex angle JPD is 320°, and the measure of angle IPJ is 105°, what is the degree measure of $\overarc{IBD}$?*

Solution: The measure of reflex angle *IPD* = measure of reflex angle *JPD* – measure of angle *IPJ* = 320° – 105° = 215°. By definition, this is also the degree measure of $\overarc{IBD}$.

MathFlash!

Whenever we need to determine the measure of an arc, we will usually specify whether the measure is in linear units or in degree form. The notation m$\overarc{AB}$ can be used to represent the degree measure of $\overarc{AB}$.

2 **Example:** *Consider the following diagram.*

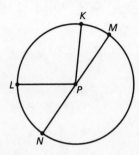

$\overline{MN}$ *is a diameter of circle P. If* $m\overset{\frown}{NK} = 155°$ *and* $m\overset{\frown}{LM} = 130°$, *what is the degree measure of* $\overset{\frown}{LK}$?

Solution: Let x represent the measure of $\overset{\frown}{LK}$.

$m\overset{\frown}{NM} = 180°$, since $\overline{NM}$ is a diameter.
Then, $m\overset{\frown}{NK} + m\overset{\frown}{LM} - x = 180°$.
By substitution, we get $155° + 130° - x = 180°$.
Simplifying, $285° - x = 180°$, and, finally, $x = 105°$.

3 **Example:** *Consider the following diagram, in which P is the center.*

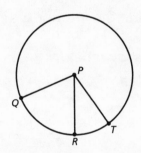

$m\angle QPR = 80°$, $m\overset{\frown}{RT} = 40°$. *What is the degree measure of the reflex* $\angle QPT$?

Solution: $m\angle RPT = m\overset{\frown}{RT} = 40°$. Since $m\angle QPT = m\angle QPR + m\angle RPT$, we can write $m\angle QPT = 80° + 40° = 120°$. Finally, the measure of reflex $\angle QPT = 360° - 120° = 240°$.

A **sector** of a circle is the region formed by two radii and an arc. The arc may be minor, major, or a semicircle.

A **segment** of a circle is the region formed by a chord and its associated minor chord. Note that this region must be less than one-half the interior of the circle. Consider Figures 19.9 and 19.10, shown below.

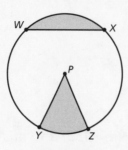

Figure 19.9

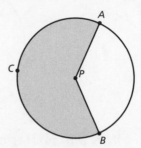

Figure 19.10

In Figure 19.9, the sector is the shaded region bounded by radii *PY* and *PZ*, and $\overset{\frown}{YZ}$. The segment is the shaded region bounded by the chord *WX* and $\overset{\frown}{WX}$.

In Figure 19.10, the sector is the shaded region bounded by radii *PA*, *PB*, and $\overset{\frown}{ABC}$, which is a major arc. In both figures, point *P* is the center.

Concentric circles have the same center, but have different sizes.

Two circles are considered **externally tangent** if they intersect at one point and their interiors do not overlap.

Two circles are **internally tangent** if they intersect at one point and the interior of one of them lies completely inside the interior of the other.

Consider Figures 19.11 and 19.12, shown below.

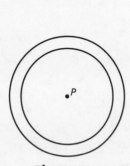

Figure 19.11

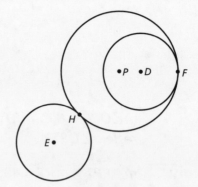

Figure 19.12

In Figure 19.11, the two circles share the same center, which is point *P*. The only way to distinguish them in a problem is to refer to the larger or the smaller circle *P*. In Figure 19.12, circles *P* and *D* are <u>internally</u> tangent at point *F*. Circles *P* and *E* are <u>externally</u> tangent at point *H*.

A polygon is considered **inscribed** in a circle if all of the vertices lie on the circle. In this situation, the circle is considered to be circumscribed about the polygon.

A polygon is considered **circumscribed** about a circle if each side of the polygon is a segment of a tangent to the circle. Consider Figures 19.13 and 19.14, shown below.

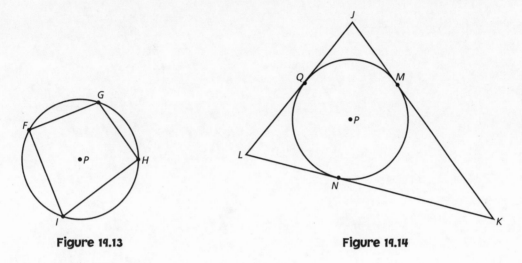

Figure 19.13   Figure 19.14

In Figure 19.13, quadrilateral *FGHI* is inscribed in circle *P*. In Figure 19.14, △*JKL* is circumscribed about circle *P*. Notice that each segment of the triangle intersects the circle exactly once. Points *M*, *N*, and *Q* are called **points of tangency**.

1.   Which one of the following represents the measure of a reflex angle?

(A)   300°   (C)   100°

(B)   175°   (D)   60°

2.   The degree measure of an arc is the same as the degree measure of _____.

(A)   the radius

(B)   a sector

(C)   its associated central angle

(D)   its associated inscribed angle

 *Test Yourself!* (continued)

3.   In which one of the following is a shaded segment being illustrated?
       (*P* is the center of each circle.)

(A)

(C)
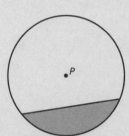

(B)
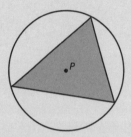

(D)

4.   Which one of the following statements is correct?

(A)   A tangent is a special type of radius.

(B)   A chord is a special type of tangent.

(C)   A diameter is a special type of chord.

(D)   A secant is a special type of diameter.

5.   Consider the diagram shown below.

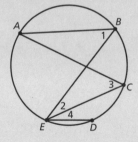

Which two inscribed angles intercept $\overset{\frown}{AE}$?

Answers: _____

6. Consider the diagram shown below, in which *P* is the center.

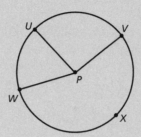

In circle *P*, if reflex angle *VPW* is 210° and *m∠UPV* = 98°, what is the degree measure of ∠*UPW*?

Answer: _____

7. Consider the diagram shown below, in which *P* is the center.

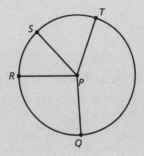

In circle *P*, if $m\overarc{TSQ} = 195°$, $m\overarc{QRS} = 112°$, and $m\overarc{RST} = 133°$, what is the degree measure of $\overarc{RS}$?

Answer: _____

8. What is the correct interpretation of two circles that are internally tangent?

    (A) They intersect at one point, and the interior of one of them lies outside the interior of the other.

    (B) They intersect at one point, and the interior of one of them lies completely inside the interior of the other.

    (C) They intersect at two distinct points.

    (D) They do not intersect.

Test Yourself! *(continued)*

9. Two concentric circles share the same _____, but have different
    _____. Which <u>two</u> of the following pairs could correctly
    fill these blanks?

    (A) center, radii   (D) center, segments

    (B) diameters, center   (E) center, circumferences

    (C) diameters, tangents

10. Consider the diagram shown below, in which *P* is the center.

    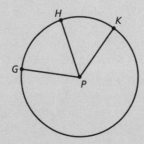

    m∠GPH = 70° and the degree measure of reflex ∠*GPK* is 225°.
    What is the degree measure of $\overset{\frown}{HK}$?

    Answer: _____

11. If each vertex of triangle *ABC* is located on circle *P*, then the triangle
    is _____ in the circle.

    Answers: _____

Circles: Formulas

In this lesson, we will explore the formulas concerning radii, diameters, circumference, arc length, sectors, and segments. These formulas will involve areas and linear lengths of various parts of a circle. You might use this lesson's formulas to find (a) the number of feet that a bicycle wheel travels in one revolution, (b) the area in square yards or feet of a circular rug, and (c) the amount of tomato sauce on a slice of pizza. Be sure that you are thoroughly familiar with all the definitions from Lesson 19.

Your Goal: When you have completed this lesson, you should be able to compute areas related to linear lengths of different parts of a circle as well as the areas of different regions.

Circles: Formulas

Look at circle *P*, shown below as Figure 20.1.

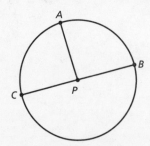

Figure 20.1

In any circle, the length of the diameter is twice the length of the radius. So, $BC = (2)(AP)$. The linear length of the entire circle, called the **circumference**, is equal to π **times the diameter**. Using *C* as the circumference and *d* as any diameter, we can write $C = \pi d$. If we use *r* as the length of the radius, we can also write $C = 2\pi r$.

In Figure 20.1, $C = (\pi)(BC) = (2\pi)(AP)$.

MathFlash!

In computational problems, unless otherwise stated, an answer may be given in units of π. We know that the approximate value of π is 3.14. However, you should try to avoid rounding off until the last step. For example, to the nearest hundredth, the value of $\dfrac{3000}{\pi}$ is 954.93, because we used the entire value of π, as kept in the memory of the calculator you are using.

If you use 3.14 as the value of π, your answer will be 955.41 (rounded off to the nearest hundredth).

1 **Example:** *What is the circumference of a circle in which the radius is 6?*

Solution: The circumference is $(2\pi)(6) = 12\pi$.

2 **Example:** *If the circumference of a circle is 20 inches, what is the diameter to the nearest hundredth?*

Solution: $20 = (\pi)(d)$, so $d = \dfrac{20}{\pi} \approx 6.37$ inches.

Look at circle Q, shown below as Figure 20.2.

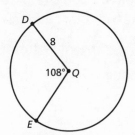

Figure 20.2

Suppose we want the linear measure of $\overset{\frown}{DE}$. We know that its degree measure is 108°. Its linear measure is found by the following formula:

$$\frac{\overset{\frown}{DE}}{\text{Circumference}} = \frac{\text{Central } \angle}{360}.$$

Let x represent $\overset{\frown}{DE}$. Then, $\dfrac{x}{(2\pi)(8)} = \dfrac{108}{360}$.

Cross-multiplying, we get $360x = 1728\pi$.

Thus, $x = \dfrac{1728\pi}{360} = 4.8\pi$.

3 **Example:**   *Look at circle P, shown below.*

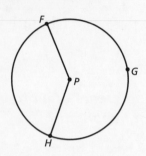

If FP = 10, and $\overarc{FGH}$ = 13π, what is the degree measure of reflex ∠FPH?

Solution: The circumference equals (2π)(10) = 20π.
Let x represent the degree measure of reflex ∠FPH.
Then, $\frac{13\pi}{20\pi} = \frac{x}{360}$. The next step is 20πx = 4680π.
Finally, $x = \frac{4680\pi}{20\pi} = 234°$.

4 **Example:**   *To the nearest hundredth, what is the circumference of a circle in which the measure of $\overarc{JK}$ is both 75° and 12π units.*

Solution: Let x represent the circumference. Then, $\frac{12\pi}{x} = \frac{75}{360}$. This leads to 75x = 4320π, so $x = \frac{4320\pi}{75} \approx 180.96$ units.

MathFlash!

It is always permissible to reduce the size of fractions when the numbers seem rather large. In Example 4, we could have reduced $\frac{75}{360}$ to $\frac{5}{24}$. The next steps would be 5x = 288π, followed by $x = \frac{288\pi}{5} \approx 180.96$.

Similar to the polygons that we studied in earlier lessons, the interior region of a circle is measured in square units. It is referred to as the **area of the circle**. The formula is given by $A = \pi r^2$, where r is the radius.

If we want the **area of a sector**, the formula we can use is $\dfrac{\textbf{Sector Area}}{\textbf{Circle Area}} = \dfrac{\textbf{Central Angle}}{\textbf{360°}}$.
This last formula can also be used if we know the area of a sector and wish to determine the measure of the central angle.

5　**Example:**　*If the radius of a circle is 15, what is the area?*

　　Solution:　$A = \pi(15)^2 = 225\,\pi$

6　**Example:**　*If the area of a circle is 26 square inches, what is the diameter to the nearest hundredth?*

　　Solution:　Since the area formula involves the radius, let x represent the radius. Then $26 = \pi\,(x)^2$. Dividing by π, $\dfrac{26}{\pi} = x^2$. So, $x = \sqrt{\dfrac{26}{\pi}} \approx 2.88$. Finally, since the radius is half the diameter, the diameter is then approximately $(2)(2.88) = 5.76$ inches.

7　**Example:**　*Consider circle Q, shown below.*

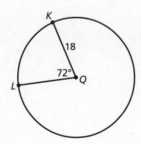

　　What is the area of the sector bounded by $\overset{\frown}{KL}$, $\overline{KQ}$, and $\overline{QL}$?

　　Solution:　The area of the circle is $(\pi)(18)^2 = 324\,\pi$. The central angle is 72°, so letting x represent the area of the sector, we can use the proportion $\dfrac{x}{324\pi} = \dfrac{72}{360}$. Then $\dfrac{x}{324\pi} = \dfrac{1}{5}$.

　　Cross-multiplying, $5x = 324\,\pi$. Thus, $x = \dfrac{324\pi}{5}$.

　　This becomes 64.8π.

8 **Example:** *Consider circle P, shown below.*

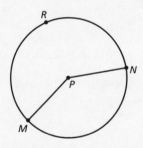

If **PM** *= 7 cm and the area of the sector bounded by* **PM**, **PN**, *and* **MRN** *is 28π cm², what is the measure of reflex ∠MPN? Round off your answer to the nearest degree.*

Solution: Let x represent the measure of reflex ∠MPN.
The area of the circle is $(\pi)(7)^2 = 49\pi$ cm².

Now, use the proportion $\dfrac{Sector\ Area}{Circle\ Area} = \dfrac{Central\ Angle}{360°}, \dfrac{28\pi}{49\pi} = \dfrac{x}{360°}$.

We can safely "drop" the "π" in the left fraction and reduce the numbers so that $\dfrac{28\pi}{49\pi} = \dfrac{4}{7}$.

Then $\dfrac{4}{7} = \dfrac{x}{360°}$ which becomes $7x = 1440°$.

Finally, $x \approx 206°$.

9 **Example:** *Consider circle Q, shown below.*

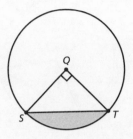

The shaded region is a segment bounded by the chord $\overline{ST}$ and the arc $\overset{\frown}{ST}$. If QS = 20, what is the area of the segment?

Solution: As yet, we have not discussed a formula for the area of a segment. By inspecting the diagram, it should seem logical that the area of this segment equals the <u>difference</u> of the area of the sector bounded by $\overline{QS}$, $\overline{QT}$, and $\overset{\frown}{ST}$, and the area of $\triangle QST$.

The area of the circle is $(\pi)(20)^2 = 400\pi$. Then letting x represent the area of the sector, and noting that the central angle is 90°, we use the proportion $\dfrac{x}{400\pi} = \dfrac{9}{360}$. Then $\dfrac{x}{400\pi} = \dfrac{1}{4}$. Cross-multiply to get $4x = 400\pi$. Thus, $x = 100\pi$.

Noting that $QT = QS = 20$, and that there is a right angle at Q, the area of the triangle is simply $\left(\dfrac{1}{2}\right)(20)(20) = 200$.

Finally, the area of the segment is $100\pi - 200$.

Example 9 illustrated how to find the area of a segment when the associated triangle is a right triangle. Do you recall how to find the area of a non-right triangle, given the lengths of all three sides?

10 **Example:** *Consider circle P, shown below.*

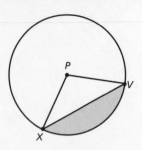

The shaded region is a segment bounded by the chord $\overline{XV}$ and the arc $\overset{\frown}{XV}$. If PX = 15, XV = 26, and m∠XPV= 120°, what is the area of the segment?

Solution: We will start with the area of the circle, which is $(\pi)(15)^2 = 225\pi$. Let x represent the area of the sector formed by $\overline{PX}$, $\overline{PV}$, and $\overset{\frown}{XV}$, and reduce $\frac{120}{360}$ to $\frac{1}{3}$.

We can use the proportion $\frac{x}{225\pi} = \frac{1}{3}$. Then, $3x = 225\pi$, so $x = 75\pi$.

Since $PV = PX = 15$, the semiperimeter of $\triangle PVX$ is $\left(\frac{1}{2}\right)(15 + 15 + 26) = 28$.

Using Heron's formula:

the area of $\triangle PVX = \sqrt{(28)(13)(13)(2)} = \sqrt{9464} \approx 97.28$.

Finally, the area of the segment is approximately $75\pi - 97.28$.

MathFlash!

In order to apply Heron's formula, you must be able to determine the lengths of all three sides.

An **annulus** (say "**ann**-yuh-luss") is defined as the region between two concentric circles, as shown below in Figure 20.3, with *P* as the center of both circles.

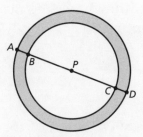

Figure 20.3

The area of the annulus is simply the difference in areas between the areas of the outermost circle and the innermost circle.

11 **Example:**   *Using Figure 20.3, suppose AD = 32 and BC = 22. What is the area of the shaded region?*

Solution: The radius of the outermost circle is $\left(\dfrac{1}{2}\right)(32) = 16$, so its area is $\pi(16^2)$. The radius of the innermost circle is $\left(\dfrac{1}{2}\right)(22) = 11$, so its area is $\pi(11^2)$. Subtract the areas of the two circles to get the area of the shaded region.
$(\pi)(16^2) - (\pi)(11^2) = 256\pi - 121\pi = 135\pi$.

12 **Example:**   *Consider the following concentric circles, with Q as the center.*

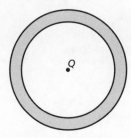

The circumference of the outermost circle is 40π, and the area of the annulus is 250π. To the nearest hundredth, what is the radius of the innermost circle?

Solution: Using $C = 2\pi r$ for the outermost circle, $40\pi = 2\pi r$. Then $r = 20$.
So the area of the outermost circle is $(\pi)(20^2) = 400\pi$.
The area of the innermost circle must be $400\pi - 250\pi = 150\pi$.
Let *x* represent the radius of the innermost circle.
Then, $(\pi)(x^2) = 150\pi$, which leads to $x^2 = 150$.
Finally, $x = \sqrt{150} \approx 12.25$.

Test Yourself!

1. To the nearest hundredth, what is the diameter of a circle whose circumference is 54 centimeters?

    Answer: _____

2. If the area of a circle is 500π, which of the following represents the value of the circumference?

    (A) 15.81π   (C) 44.72π

    (B) 22.36π   (D) 79.58π

3. Consider circle *P*, shown below.

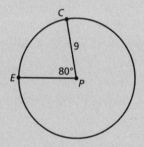

    In terms of π, what is the linear measure of $\overset{\frown}{CE}$?

    Answer: _____

4. Consider circle *Q*, shown below.

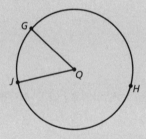

    If *GQ* = 16 and $\overset{\frown}{GHJ}$ = 28π, what is the degree measure of reflex ∠*GQJ*?

    Answer: _____

Test Yourself! (continued)

5. Consider circle *P*, shown below.

    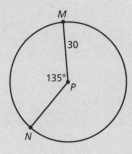

    Which of the following is <u>closest</u> to the area of the sector bounded by $\overline{PM}$, $\overline{PN}$, and $\overarc{MN}$?

    (A) 308π   (C) 328π

    (B) 318π   (D) 338π

6. Given two concentric circles in which their radii are 25 centimeters and 40 centimeters, what is the area of the associated annulus? You may leave your answer in terms of π.

    Answer: _____

7. Consider circle *Q*, shown below.

    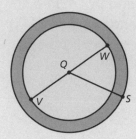

    QS = 12 and the area of the shaded region is 56π. What is the <u>best</u> approximation for the value of $\overline{VW}$?

    (A) 18.76   (C) 9.38

    (B) 15.88   (D) 7.94

 Test Yourself! (continued)

8.   Consider circle *P*, shown below.

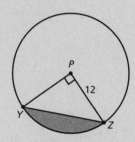

What is the area of the shaded portion? You may leave your answer in terms of π.

Answer: _____

9.   Consider circle *Q*, shown below.

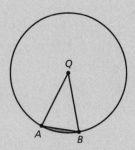

If *QA* = 32, *AB* = 22, and *m∠AQB* = 40°, to the nearest integer, what is the area of the shaded segment?

(A)   30   (C)   24

(B)   27   (D)   21

10.   The area of a circle is 81π. If the area of a particular sector of this circle is 31π, what is the central angle formed by this sector to the nearest degree?

(A)   150°   (C)   142°

(B)   146°   (D)   138°

Circles: Angle Theorems

In this lesson, we will explore theorems concerning chords, tangents, secants, and angles associated with circles. We will discover how the angle measures are determined from the various ways in which they are formed with line segments related to a circle.

Your Goal: When you have completed this lesson, you should be able to compute the measures of angles that result from the various ways in which they appear either inside, on, or outside the circle.

Circles: Angle Theorems

Look at circles P and Q, shown below as Figures 21.1 and 21.2.

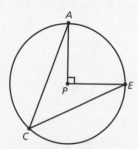

Figure 21.1

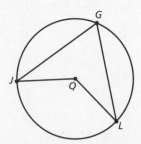

Figure 21.2

In Figure 21.1, $\angle ACE$ is an inscribed angle, and $\angle APE$ is a central angle. Each of these angles intercepts $\overparen{AE}$, whose degree measure is 90°. $m\angle APE = 90°$, since the measure of a central angle is always equal to the measure of its intercepted arc. Since this diagram is drawn to scale, use your protractor to measure $\angle ACE$. You should discover that its measure is 45°, which is one-half of 90°. In Figure 21.2, each of $\angle JGL$ and $\angle JQL$ intercepts $\overparen{JL}$. In using your protractor, you will find that $m\angle JQL = 130°$ and $m\angle JGL = 65°$. This information suggests the following theorem.

THEOREM 1   In a circle, the measure of an inscribed angle is one-half the degree measure of its intercepted arc.

Let's look at a situation involving an angle formed by a tangent and a chord, as shown below in Figures 21.3 and 21.4.

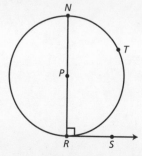

Figure 21.3

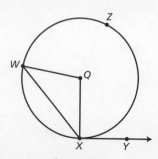

Figure 21.4

In Figure 21.3, $\overline{NR}$ is a diameter of circle P, and $\overrightarrow{RS}$ is a tangent ray to this circle. $\overline{NR}$ is perpendicular to $\overrightarrow{RS}$, $m\angle NRS = 90°$. You can clearly see that $m\overarc{NTR} = 180°$, since $\overarc{NTR}$ is a semicircle. Notice that $\overarc{NTR}$ is the intercepted arc of $\angle NRS$, and $m\angle NRS$ is one-half of $m\overarc{NTR}$.

In Figure 21.4, $\overrightarrow{XY}$ is a tangent ray to the circle. This figure is also drawn to scale, with $m\angle WQX = 100°$ and $m\angle WXY = 130°$. We know that $m\overarc{WX} = 100°$, and since the degree measure of a complete circle is 360°, we know that $m\overarc{WZX} = 360° - 100° = 260°$. Notice that $\overarc{WZX}$ is the intercepted arc of $\angle WXY$, and that $m\angle WXY$ is one-half of $m\overarc{WZX}$. This information suggests the following theorem.

THEOREM 2   In a circle, the measure of an angle formed by a tangent ray and a chord is one-half the measure of its intercepted arc.

In Figure 21.4, $\triangle WQX$ is isosceles, because it consists of two radii. You should have no difficulty in determining that the measure of each of $\angle QWX$ and $\angle QXW$ is 40°. We were already given that $m\angle WXY = 130°$. Furthermore, this implies that $m\angle QXY = 130° - 40° = 90°$. This conclusion leads to our next theorem.

THEOREM 3   In a circle, the measure of an angle formed by a tangent and a radius (or diameter) is always 90°. (Incidentally, "tangent ray" may be substituted for "tangent.")

MathFlash!

Looking back at Figure 21.3, once we know that $\overrightarrow{RS}$ is tangent to circle P and that $\overline{NR}$ is a diameter, the measure of $\angle NRS$ is automatically 90°.

1 **Example:**   *Consider circle P, shown below.*

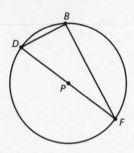

If the measure of $\overset{\frown}{BF}$ is 96° larger than the measure of $\overset{\frown}{BD}$, what is the measure of ∠BFD?

Solution: First, note that $m\overset{\frown}{DBF} = 180°$, since this arc represents a semicircle. Let x represent the measure of $\overset{\frown}{BD}$, and let $x + 96$ represent $\overset{\frown}{BF}$. Then write $x + (x + 96) = 180$.
The next steps are $2x + 96 = 180$, followed by $2x = 84$, so $x = 42°$. Since ∠BFD is an inscribed angle with $\overset{\frown}{BD}$ as its intercepted arc,
$$m∠BFD = \left(\frac{1}{2}\right)(42) = 21°.$$

2 **Example:**   *Consider circle Q, shown below.*

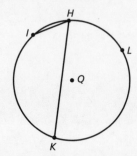

If the measure of $\overset{\frown}{IK}$ is three times as large as the measure of $\overset{\frown}{HI}$, and $m\overset{\frown}{HLK} = 220°$, what is the measure of ∠IHK?

Solution: Let x represent the measure of $\overset{\frown}{HI}$, and let $3x$ represent the measure of $\overset{\frown}{IK}$. So we can write $x + 3x + 220 = 360$. The next steps are $4x + 220 = 360$, followed by $4x = 140$, so $x = 35°$.

Then, $3x = m\overset{\frown}{IK} = 105°$. Finally, $m ∠IHK = \left(\frac{1}{2}\right)(m\overset{\frown}{IK}) = 52.5°$.

We will explore the relationship between the angle formed by the tangent rays and the measures of the two intercepted arcs.

Let's consider the situation that involves two tangent rays to a given circle from a common point. This is illustrated in Figure 21.5, shown below.

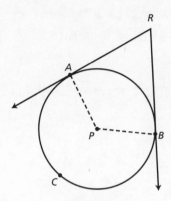

Figure 21.5

Figure 21.5 has been drawn to scale, so that $m\angle APB = 125°$, where P is the center. Since $\angle APB$ is a central angle, $m\overset{\frown}{AB} = 125°$ and $m\overset{\frown}{ACB} = 360 - 125 = 235°$. Use your protractor to measure $\angle ARB$. Your measurement should read 55°.

We are looking for a connection among the measurements 125°, 235°, and 55°. The relationship among these numbers, although not obvious, really does exist. If we subtract 125° from 235°, then divide this difference by 2; the result is 55°.

Look at Figure 21.6, shown below, in which Q is the center of the circle and $m\angle DQE = 40°$.

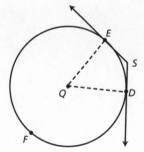

Figure 21.6

This means that $m\overset{\frown}{ED} = 40°$ and $m\overset{\frown}{EFD} = 360° - 40° = 320°$. Without using your protractor, and using the fact that each of $\angle QDS$ and $\angle QES$ is a right angle, $m\angle ESD = 360 - 40 - 90 - 90 = 140°$. This is true because the sum of angles of a quadrilateral must be 360°. Let's use the same computational "trick" we used for Figure 21.5. We subtract 40° from 320° and then divide this difference by 2. The result is 140°, which is exactly the measure of $\angle ESD$.

The information we received from Figures 21.5 and 21.6 suggests the following theorem.

THEOREM 4 The measure of an angle formed by two tangents to a circle is equal to one half the difference of the measures of the intercepted arcs.

We will now consider an angle formed by a tangent and a secant, drawn to a circle from an exterior point. This is illustrated in Figure 21.7, shown below:

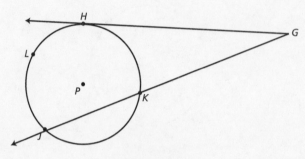

Figure 21.7

Figure 21.7 has been drawn to scale so that $m\overset{\frown}{HK} = 90°$ and $m\overset{\frown}{HLJ} = 150°$. Using your protractor to measure $\angle HGK$, you should discover that $m\angle HGK = 30°$.

Notice that $30° = \left(\dfrac{1}{2}\right)(150 - 90)$.

So we can now see that **the measure of the angle formed by the tangent and secant is one-half the difference of the two intercepted arcs.**

Let's illustrate two examples of an angle formed by two secants, as shown below in Figures 21.8 and 21.9.

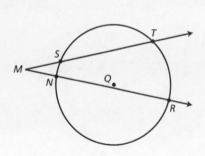

Figure 21.8

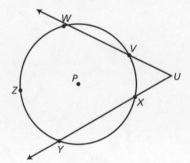

Figure 21.9

Points Q and P represent the centers of the circles in Figures 21.8 and 21.9, respectively. These figures are drawn to scale. In Figure 21.8, $m\overset{\frown}{RT} = 65°$ and $m\overset{\frown}{SN} = 25°$.
Use your protractor to verify that $m\angle NMS = 20°$.
In Figure 21.9, $m\overset{\frown}{WZY} = 145°$ and $m\overset{\frown}{VX} = 45°$.
Use your protractor to verify that $m\angle VUX = 50°$.
For each of these figures, note that the measure of the angle formed by the two secants is exactly one-half of the difference of the measures of their intercepted arcs.
Specifically, $20 = \left(\frac{1}{2}\right)(65 - 25)$ and $50 = \left(\frac{1}{2}\right)(145 - 45)$.

Then $m\angle TMR = \left(\frac{1}{2}\right)(\overset{\frown}{TR} - \overset{\frown}{SN})$ and $m\angle WUY = \left(\frac{1}{2}\right)(\overset{\frown}{WY} - \overset{\frown}{VX})$.

Let's return to Theorem 4 and rewrite it to be more inclusive.

(NEW) THEOREM 4 The measure of an angle formed by (a) two tangents, (b) a tangent and a secant, or (c) two secants to a circle from a point outside the circle is equal to one-half the difference of the intercepted arcs.

MathFlash!

Remember that the measure of an arc is always equal to the measure of its corresponding central angle. For example, if you wanted to verify that the measure of $\overset{\frown}{NS}$ is 25°, simply draw the radii $\overline{QS}$ and $\overline{QN}$. Then measure $\angle NQS$.

3 **Example:** *Consider circle P, shown below.*

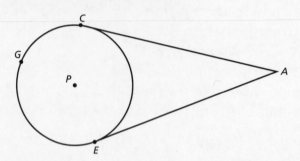

$\overline{AC}$ and $\overline{AE}$ are tangent segments. If $m\widehat{CE} = 156°$, what is the measure of $\angle CAE$?

Solution: Let x represent the measure of $\angle CAE$. Then, since $m\widehat{CGE} = 360 - 156 = 204°$, we can write

$$x = \left(\frac{1}{2}\right)(204 - 156) = \left(\frac{1}{2}\right)(48) = 24°.$$

4 **Example:** *Consider circle Q, shown below.*

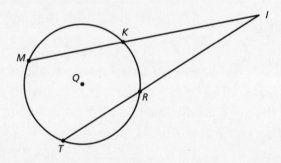

$\overline{IM}$ and $\overline{IT}$ are secant rays. If $m\angle KIR = 18°$ and $m\widehat{MT} = 81°$, what is the measure of $\widehat{KR}$?

Solution: Let x represent the measure of $\widehat{KR}$. Then, $18 = \left(\frac{1}{2}\right)(81 - x)$.

Simplify, $18 = 40.5 - \frac{1}{2}x$.

Next, $-22.5 = -\frac{1}{2}x$.

Finally, $x = 45°$.

5 **Example:** *Consider circle P, shown below.*

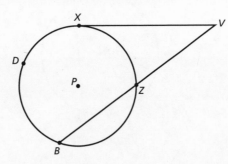

*XV is a tangent segment, and BV is a secant segment. If
$m\angle XVZ = 42°$ and $m\overset{\frown}{XZ} = 90°$, what is the measure of $\overset{\frown}{BZ}$?*

Solution: Careful! Even though it appears that $\overset{\frown}{BX}$ is a semicircle, this would
be wrong. Let x represent the measure of $\overset{\frown}{BDX}$.

Write $42 = \left(\dfrac{1}{2}\right)(x - 90)$.

Simplify, $42 = \dfrac{1}{2}x - 45$.

Next, $87 = \dfrac{1}{2}x$. Then $x = 174°$.

Don't stop yet! We need to find the measure of $\overset{\frown}{BZ}$. Since the sum
of the measures of $\overset{\frown}{BDX}$, $\overset{\frown}{XZ}$, and $\overset{\frown}{BZ}$ must be 360°,
$m\overset{\frown}{BZ} = 360 - 174 - 90 = 96°$.

MathFlash!

*We began the solution by letting x represent the measure of $\overset{\frown}{BDX}$.
Even though we discovered that $\overset{\frown}{BDX}$ is actually a minor arc, it was
still permissible to name it using three letters. Whenever an arc is
actually known to be a **major** arc, it <u>must</u> either be named using
three letters or it <u>must</u> be identified as a major arc. The first of
these two options is the one used most often.*

Our final situation involving angles for this lesson will concern angles formed by a pair
of intersecting chords. If the chords are both diameters, then we simply have central
angles. So, for each of the following examples, at least one of the chords will not be a
diameter. Consider Figure 21.10, shown below.

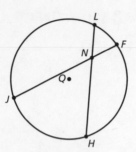

Figure 21.10

In circle Q, chords $\overline{FJ}$ and $\overline{HL}$ intersect at point N. The figure has been drawn to scale so that $m\widehat{LF} = 32°$ and $m\widehat{HJ} = 86°$. Use your protractor to verify that the measure of $\angle LNF$ is 59°. In similar fashion, given that $m\widehat{FH} = 124°$ and $m\widehat{JL} = 118°$, use your protractor to verify that the measure of $\angle LNJ$ is 121°. (Of course, you recognize that $\angle LNF$ and $\angle LNJ$ are adjacent supplementary angles, so the sum of their measures must be 180°.)

Take a closer look at the measures of $\widehat{FH}$, $\widehat{JL}$, and $\angle LNJ$. Can you spot a relationship among the numbers 124, 118, and 121? If you guessed that 121 is the arithmetic average of 118 and 124, you are right on target! Likewise, look at the measures of $\widehat{LF}$, $\widehat{HJ}$, and $\angle LNF$. The three numbers under consideration are 32, 86, and 59. Sure enough, one of these numbers, namely, 59, is the arithmetic average of the other two numbers.

In case you are not sure of the meaning of "the arithmetic average of two numbers," it is found by adding the numbers and then dividing by 2. For example, the arithmetic average of 32 and 86 is found by calculating $\dfrac{32 + 86}{2} = \dfrac{118}{2} = 59$.

Before we write the associated theorem for Figure 21.10, we will show another example in which a diameter is one of the chords. Consider Figure 21.11, shown below.

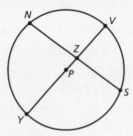

Figure 21.11

In circle P, chords $\overline{VY}$ and $\overline{NS}$ intersect at point Z. Notice that $\overline{VY}$ is also a diameter. The figure has been drawn to scale so that $m\widehat{NV} = 80°$ and $m\widehat{SY} = 116°$. Use your protractor to verify that $m\angle NZV = 98°$. Also, $m\widehat{NY} = 100°$ and $m\widehat{SV} = 64°$. Again, using your protractor, you can verify that $m\angle NZY = 82°$. In each case, notice that the measure of an angle formed by the two chords is the arithmetic average of two numbers that represent arc measures.

THEOREM 5 The measure of an angle formed by two intersecting chords is equal to the arithmetic average of the measures of the arc it intercepts and the arc intercepted by its vertical angle.

Another name for "vertical" angles is "opposite" angles. Vertical angles always have equal measures. Just as we verified that $m\angle NZV = 98°$, we would also expect that $m\angle SZY = 98°$. Likewise, we expect that $m\angle SZV = 82°$ because $180 - 98 = 82$.

6 **Example:** Consider circle Q, shown below.

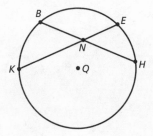

Chords $\overline{BH}$ and $\overline{EK}$ intersect at point N. If $m\overset{\frown}{EH} = 35°$ and $m\overset{\frown}{BK} = 47°$, what is the measure of $\angle ENH$?

Solution: Let x represent $m\angle ENH$. Then, $x = \left(\dfrac{1}{2}\right)(35 + 47) = \left(\dfrac{1}{2}\right)(82) = 41°$.

7 **Example:** *Consider circle P, shown below.*

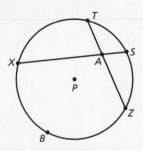

Chords $\overline{TZ}$ and $\overline{SX}$ intersect at point A. If m∠SAT = 108° and m$\overarc{ZBX}$ = 166°, what is the measure of $\overarc{ST}$?

Solution: Let x represent $m\overarc{ST}$. Then, $108 = \left(\dfrac{1}{2}\right)(x + 166)$.

Simplify this equation to $108 = \dfrac{1}{2}x + 83$.

The next two steps are $25 = \dfrac{1}{2}x$, followed by $x = 50°$.

Our final theorem for this lesson will involve an inscribed quadrilateral in circle Q, as shown below in Figure 21.12.

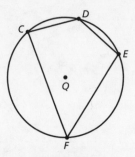

Figure 21.12

CDEF is an inscribed quadrilateral. We already know that the sum of the measures of all four (inscribed) angles is 360°. Using your protractor for this particular quadrilateral, you can determine that $m\angle C = 81°$, $m\angle D = 116°$, $m\angle E = 99°$, and $m\angle F = 64°$.
Can you spot a relationship that exists for each of two pairs of these angles?
If you focus on the measures of $\angle C$ and $\angle E$, you may quickly notice that they represent supplementary angles. Now, if you focus on the measures of $\angle D$ and $\angle F$, you will again notice that their sum is 180°.

THEOREM 6 Given an inscribed quadrilateral of a circle, opposite angles are supplementary.

8 **Example:** *Consider circle P, with inscribed quadrilateral GHIJ, as shown below.*

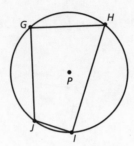

If m∠H = 72° and m∠I = 94°, what are the measures of ∠G and ∠J?

Solution: This will be quick! $m\angle G = 180 - 94 = 86°$ and $m\angle J = 180 - 72 = 108°$.

9 **Example:** *Using the diagram in Example 8, suppose it is given that $m\widehat{GH} = 90°$. What is the measure of $\widehat{GJ}$?*

Solution: First notice that $m\widehat{GJ} + m\widehat{GH} = m\widehat{JGH}$.
Now, see that the inscribed $\angle I$ intercepts $\widehat{JGH}$.
Since $m\angle I = 94 = \left(\dfrac{1}{2}\right)(m\widehat{JGH})$,
doubling 94 means that $m\widehat{JGH} = 188°$.
Letting x represent the measure of $\widehat{GJ}$, $x + 90° = 188°$.
Finally, $x = 98°$.

MathFlash!

In the solution of Example 9, we used the material that you learned in the beginning of this lesson. It never hurts to reinforce the earlier concepts!

10 **Example:**   Consider circle Q, with inscribed quadrilateral KLMN, as shown below.

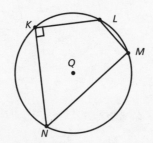

If $m\widehat{MN}$ = 145° and $m\widehat{KL}$ = 75°, what is the measure of ∠N?

Solution: Since $m\angle K$ = 90°, we know that $m\widehat{LMN}$ = 180°.
Then $m\widehat{LM}$ = 180° − 145° = 35°.
This means that $m\widehat{KLM}$ = 75 + 35 = 110°.
Finally, $m\angle N = \left(\dfrac{1}{2}\right)(110) = 55°$.

Test Yourself!

1.   Look at circle *P*, shown below.

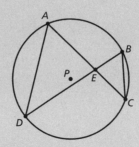

Which angle must have the same measure as ∠A?

Answer: _____

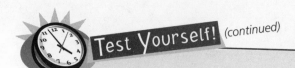

2.   Look at circle *Q*, shown below, in which $\overrightarrow{FG}$ is a tangent ray.

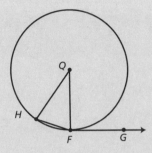

If *m∠HQF* = 54°, what is the measure of ∠*HFG*?

Answer: _____

3.   Look at circle *P*, shown below.

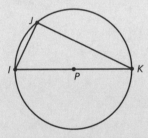

If the measure of $\widehat{JK}$ is four times as large as the measure of $\widehat{IJ}$, what is the measure of ∠*I*?

Answer: _____

4.   Look at circle *Q*, shown below.

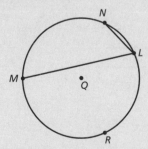

If *m∠L* = 44° and the measure of $\widehat{MRL}$ is 100° larger than the measure of $\widehat{LN}$, what is the measure of $\widehat{MRL}$?

Answer: _____

Test Yourself! (continued)

5.   Look at circle *P*, shown below, in which $\overrightarrow{ST}$ and $\overrightarrow{SU}$ are tangent rays.

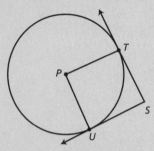

How many right angles **must** *PTSU* contain?

Answer: _____

6.   Look at circle *Q*, shown below, in which $\overrightarrow{VW}$ is a tangent ray and $\overrightarrow{VY}$ is a secant ray.

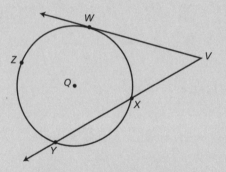

If $m\angle V = 26°$ and $m\overparen{WX} = 77°$, what is the measure of $\overparen{WZY}$?

Answer: _____

Test Yourself! (continued)

For questions 7 and 8, use circle *P*, shown below. $\overline{AC}$ and $\overline{BD}$ are chords that intersect at point *E*.

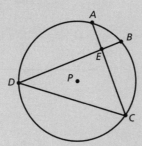

7. If $m\widehat{DAB} = 150°$ and $m\angle C = 62°$, what is the measure of $\widehat{AB}$?

    Answer: _____

8. Using the information in question 7, combined with $m\widehat{DC} = 104°$, what is the measure of $\angle DEC$?

    Answer: _____

For questions 9 and 10, use circle *Q*, shown below. *FHJL* is an inscribed quadrilateral.

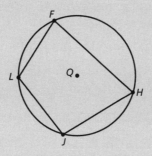

9. If $m\widehat{FL} = 70°$ and $m\widehat{JL} = 76°$, what is the measure of $\angle L$?

    Answer: _____

10. Using the information in question 9, combined with the fact that the measure of $\angle J$ is 22° larger than the measure of $\angle F$, what is the measure of $\widehat{LFH}$?

    Answer: _____

Circles: Segment Theorems

In this lesson, we will continue to explore theorems concerning parts of a circle. In Lesson 21, our focus was on the angles formed by chords, tangents, and secants. In this lesson, our objective will be to determine the relationships among line segments within and outside a circle.

Your Goal: When you have completed this lesson, you should be able to compute the lengths of chords, tangent segments, secant segments, and portions of these segments.

LESSON 22

Circles: Segment Theorems

1 **Example:** *We will begin with a single tangent to a circle, as shown below.*

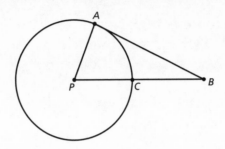

$\overline{AB}$ *is a tangent segment to circle P. If PC = 5 and AB = 12, what is the value of PB?*

Solution: Let x represent *PB*. We recall that a tangent (or a tangent segment) is perpendicular to a circle at the point of tangency, so that *ABP* is a right triangle. The next step is to recognize that $PA = PC = 5$. Hopefully, you remember our friend Pythagoras. By the Pythagorean theorem, $5^2 + 12^2 = x^2$.

The left side simplifies to 169, so $x = \sqrt{169} = 13$.

MathFlash!

If the square root value is not exact, we will simply round off to the nearest hundredth. Notice that this method would have been used if we wanted the value of BC.

In this example, $BC = PB - PC$, $13 - 5 = 8$.

2   Example: *Consider circle Q, with tangent segment $\overline{DE}$, as shown below.*

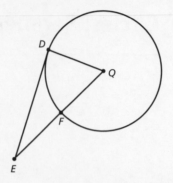

If EF = 9 and DQ = 7, what is the value of DE?

Solution: Let x represent *DE*. Since *FQ = DQ* = 7, we know that *EQ* = 9 + 7 = 16. By the Pythagorean theorem, $x^2 + 7^2 = 16^2$. Then, $x^2 + 49 = 256$. The last two steps are $x^2 = 207$ and $x = \sqrt{207} \approx 14.39$.

Let's look at a situation involving **two tangent segments to a circle** from a given exterior point, as shown below in Figure 22.1.

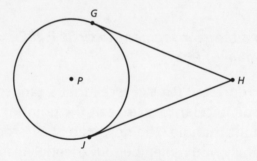

Figure 22.1

If you take your ruler and measure *GH* and *HJ*, each of which is a tangent segment to circle *P*, you will find them to be equal in length. This information leads to the following theorem.

THEOREM 1   The lengths of two tangent segments from an exterior point of a circle are equal.

Now we will look at a situation involving **a tangent segment and a secant segment, drawn from an exterior point to a circle**. This is shown below in Figure 22.2.

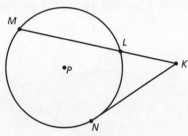

Figure 22.2

$\overline{KN}$ is the tangent segment and $\overline{KM}$ is the secant segment to circle P. Since this figure is drawn to scale, use your ruler to measure $\overline{KM}$, $\overline{KL}$, and $\overline{KN}$. You will discover that $KL \approx \frac{5}{8}$ inches, $KM \approx 1\frac{3}{4}$ inches, and $KN \approx 1\frac{1}{16}$ inches. The relationship among these three numbers will not be immediately clear. In fact, even if you spent a long time, it would be difficult to reach any conclusion. If we multiply KL by KM, we get $\left(\frac{5}{8}\right)\left(\frac{7}{4}\right) = \frac{35}{32} = 1.09375$. Let's convert $1\frac{1}{16}$ to its decimal equivalent of 1.0625. If we square 1.0625, we get about 1.1289.

Normally, we would not compute decimal values to this many places, but you can probably see why we went "the extra mile." The difference between 1.1289 and 1.09375 is about 0.035, so we would be tempted to say that they are nearly equal. This would imply that $(KN)^2 = (KM)(KL)$, which leads to the following theorem.

THEOREM 2   If a tangent segment and a secant segment are drawn to a circle from an exterior point, then the square of the length of the tangent segment is equal to the product of the length of the secant segment and the portion of the secant segment that lies outside the circle.

3 **Example:**   *Consider circle Q, with tangent segment $\overline{RS}$ and secant segment $\overline{RU}$, as shown below.*

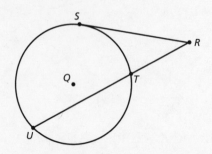

If RU = 30 and RT = 21, what is the value of RS?

Solution: Using Theorem 2, $(RU)(RT) = (RS)^2$. Letting x represent RS, we can substitute the known values. $(30)(21) = x^2$. Then, $x = \sqrt{630} \approx 25.1$.

4 **Example:**   *Consider circle P, with tangent segment $\overline{VY}$ and secant segment $\overline{VX}$, as shown below.*

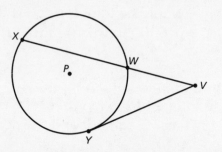

If VW = 8 and VY = 16, what is the value of WX?

Solution: Let x represent WX, so that $x + 8$ represents VX.
Since $(VX)(VW) = (VY)^2$, $(x + 8)(8) = 16^2$.
Then, $8x + 64 = 256$, followed by $8x = 192$ and $x = 24$.

Our next objective is to evaluate the situation involving **any two secants to a circle, drawn from an exterior point**. This is shown below in Figure 22.3.

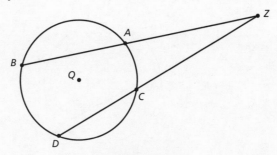

Figure 22.3

$\overline{ZB}$ and $\overline{ZD}$ are secant segments to circle Q. This figure has been drawn to scale so that you can measure $\overline{ZA}$, $\overline{ZB}$, $\overline{ZC}$, and $\overline{ZD}$. You will find that $ZA \approx 1\frac{7}{16}$ inches, $ZB \approx 2\frac{5}{8}$ inches, $ZC \approx 1\frac{1}{2}$ inches, and $ZD \approx 2\frac{1}{2}$ inches. Based on what you learned from Theorem 2, you can probably take an educated guess concerning how these numbers are related. Calculate $(ZA)(ZB) = \frac{23}{16} \times \frac{21}{8} = \frac{483}{128} \approx 3.77$. Now calculate $(ZC)(ZD) = \frac{3}{2} \times \frac{5}{2} = \frac{15}{4} = 3.75$. The difference in these two products is about 0.02, which leads us to the following theorem.

THEOREM 3   Two secant segments to a circle are drawn from the same exterior point. The product of one secant segment and its corresponding exterior portion equals the product of the other secant segment and its corresponding portion.

5   **Example:**   *Consider circle P, with secant segments $\overline{EG}$ and $\overline{EJ}$, as shown below.*

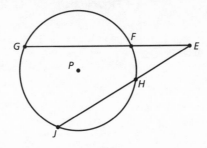

If EG = 25, EF = 10, and EH = 15, what is the value of EJ?

Solution: Based on Theorem 3, $(EG)(EF) = (EJ)(EH)$. Let x represent EJ. Then, $(25)(10) = (x)(15)$. Thus, $15x = 250$. This leads to the answer of $x = 16.\overline{6}$.

Example: *Consider circle Q, with secant segments $\overline{LN}$ and $\overline{LS}$, as shown below.*

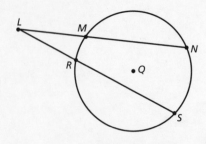

If LM = 6, MN = 10, and LR = 5, what is the value of RS?

Solution: Let *x* represent *RS*, so that *x* + 5 represents *LS*.
Our basic equation will be (*LN*)(*LM*) = (*LS*)(*LR*).
We can easily determine that *LN* = 16, since *LN* = *LM* + *MN*.
By substitution, (16)(6) = (*x* + 5)(5).
Then, 96 = 5*x* + 25, followed by 5*x* = 71, and finally *x* = 14.2.

Our last objective is to examine the situation when **two chords, not both diameters, intersect in a circle**. This implies that the point of intersection is <u>not</u> the center.

This is shown below in Figure 22.4.

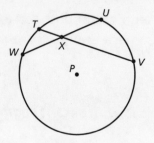

Figure 22.4

$\overline{TV}$ and $\overline{UW}$ are chords of circle P that intersect at point X. This figure has been drawn to scale so that you can measure $\overline{TX}$, $\overline{VX}$, $\overline{UX}$, and $\overline{WX}$. You should find that $TX \approx \frac{1}{4}$ inch, $VX \approx \frac{3}{4}$ inch, $UX \approx \frac{7}{16}$ inch, and $WX \approx \frac{7}{16}$. The connection among these numbers resembles the connection we discovered with the secants. We find that $(TX)(VX) = \frac{3}{16} = .1875$, and $(UX)(WX) = \frac{49}{256} \approx .1914$. The difference between these two decimal numbers is less than 0.004. Here comes the theorem!

THEOREM 4   Two chords intersect in a circle. The product of the two resulting segments of one chord equals the product of the two resulting segments of the other chord.

7   **Example:**   *Consider circle Q, with chords $\overline{AB}$ and $\overline{YZ}$ intersecting at point C.*

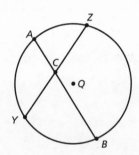

If AC = 7, BC = 24, and CY = 13, what is the value of CZ?

Solution: Theorem 4 gives us the basic equation $(AC)(BC) = (CY)(CZ)$.
Letting x represent CZ, $(7)(24) = (13)(x)$.
Then $168 = 13x$, and $x \approx 12.92$.

8 **Example:** *Consider circle P, with chords $\overline{DF}$ and $\overline{EG}$ intersecting at point H.*

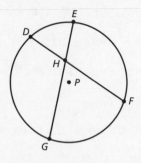

If DH = 18, DF = 38, and GH is twice as large as EH, what is the value of EH?

Solution: Let x represent *EH*, and let $2x$ represent *GH*. Note that *HF* = *DF − DH* = 20. The basic equation is (*EH*)(*GH*) = (*DH*)(*HF*). By substitution, we get $(x)(2x) = (18)(20)$. Then, $2x^2 = 360$. The next two steps are $x^2 = 180$, followed by $x = \sqrt{180} \approx 13.42$.

9 **Example:** *Consider circle Q, with diameter $\overline{JN}$ and chord $\overline{KL}$ intersecting at point M.*

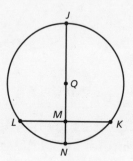

If JQ = 20, QM = 15, and LM = MK, what is the value of KL?

Solution: Let x represent each of *LM* and *MK*. If *JQ* = 20, then *QN* =20, since each one represents a radius. *MN* = *QN − QM* = 5. Also, we can write *JM* = *JQ + QM* = 35. Now, using Theorem 4, $(x)(x) = (35)(5)$. This equation simplifies to $x^2 = 175$, which means that $x = \sqrt{175} \approx 13.23$. Finally, *KL* = (2)(13.23) = 26.46.

1.   In circle *P*, shown below, $\overline{AB}$ and $\overline{AC}$ are tangent segments.

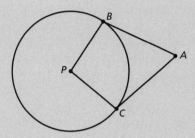

Which one of the following statements <u>must</u> be true?

(A)   *AB* = *AC* = *PB* = *PC*

(B)   No two of *AB*, *AC*, *PB*, and *PC* are equal.

(C)   *AB* = *AC* and *PB* = *PC*

(D)   *AB* = *PB* and *AC* = *PC*

2.   In circle *Q*, shown below, $\overline{DE}$ is a tangent segment.

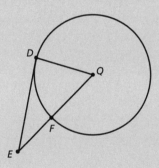

If *ED* = 30 and *DQ* = 16, what is the value of *EF*?

Answer: _____

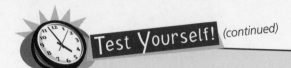 Test Yourself! (continued)

3.   In circle *P*, shown below, $\overline{GH}$ is a tangent segment.

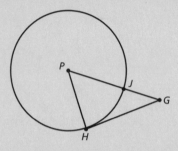

If *PH* = 10 and *GH* = 12, what is the value of *JG* to the nearest hundredth?

Answer: _____

4.   In circle *Q*, shown below, $\overline{KM}$ is a secant segment and $\overline{KN}$ is a tangent segment.

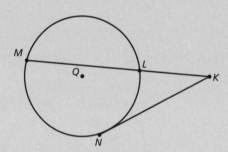

If *KL* = 13 and *ML* = 19, then which of the following is the best approximation for *KN*?

(A)   15.7

(C)   20.4

(B)   18.1

(D)   22.8

5. In circle *P*, shown below, $\overline{RU}$ is a tangent segment and $\overline{RT}$ is a secant segment.

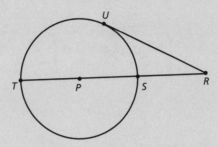

If *RU* = 24 inches and *RT* = 32 inches, how many inches is the diameter of the circle?

Answer: _____

6. In circle *Q*, shown below, $\overline{VX}$ and $\overline{VZ}$ are secant segments.

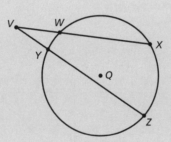

If *VX* = 18, *VW* = 5, and *VZ* = four times as large as *VY*, what is the value of *VY* to the nearest hundredth?

Answer: _____

7.   In circle *P*, shown below, *AE* and *AG* are secant segments.

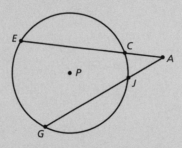

If *AC* = 8, *EC* = 20, and *AG* = 25, to the nearest integer, which one of the following is the value of *AJ*?

(A)   9   (C)   11

(B)   10   (D)   12

8.   In circle Q, shown below, $\overline{LT}$ is a diameter, $\overline{NR}$ is a chord, and *W* is the point of intersection.

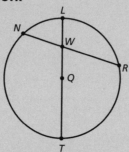

If *NW* = 3, *LW* = 2.5, and *RW* = 5, what is the length of the diameter?

Answer: _____

9. In circle *P*, shown below, diameter $\overline{FH}$ intersects chord $\overline{BD}$ at point *J*.

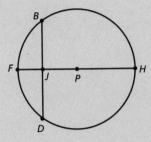

If *FJ* = 2.5, *JH* = 16, and *BJ* = *JD*, what is the value of *BJ* to the nearest hundredth?

Answer: _____

10. In circle *Q*, shown below, $\overline{SV}$ and $\overline{KM}$ are chords that intersect at point *Z*.

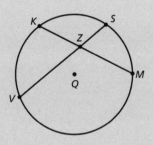

If *SV* = 25, *SZ* = 5, and *KZ* = 8, what is the value of *MZ*?

Answer: _____

QUIZ FIVE

LESSONS 19-22

1. In circle *P*, shown below, *BEGJ* is an inscribed quadrilateral.

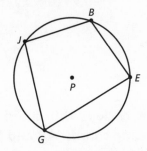

If *m∠G* = 72° and *m∠E* = 84°, what is the measure of $\overset{\frown}{BEG}$?

A 186°

B 192°

C 198°

D 204°

2. In circle *Q*, with a radius of 9, the linear measure of $\overset{\frown}{MN}$ is 7π. What is the measure of *∠MQN*?

A 280°

B 240°

C 200°

D 140°

3. A diameter is considered the longest _____ in a circle.

A secant

B chord

C tangent

D arc

4. Consider circle *Q*, with $\overrightarrow{RV}$ and $\overrightarrow{RX}$, as shown below.

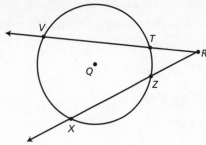

If *m∠R* = 32° and $m\overset{\frown}{VX}$ = 102°, what is the measure of $\overset{\frown}{TVZ}$?

A 322°

B 300°

C 254°

D 226°

5. Consider circle *P*, with diameter $\overline{CE}$, as shown below.

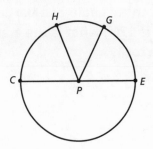

If $m\overset{\frown}{CG}$ = 133° and $m\overset{\frown}{EH}$ = 109°, what is the measure of $\overset{\frown}{GH}$?

A 24°

B 47°

C 62°

D 71°

6. Consider circle Q, with intersecting chords $\overline{NR}$ and $\overline{KM}$, as shown below.

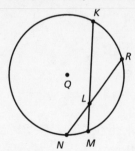

If LR = 8 inches, ML = 6 inches, and KL = 3 inches greater than LN, what is the value of NR?

A 17 inches

B 15 inches

C 11 inches

D 9 inches

7. Two concentric circles intersect in how many points?

A 0

B 1

C 2

D 3

8. The circumference of a circle is 60π. If the central angle of a particular sector is 150°, what is the area of this sector?

A 125π

B 200π

C 250π

D 375π

9. In circle Q, shown below, $\overline{ST}$ is a tangent segment and $\overline{SV}$ is a secant segment.

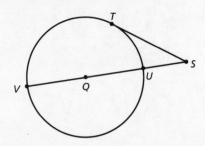

If the radius of this circle is 10 and SV = 36, what is the value of ST?

A 22

B 24

C 26

D 28

10. Consider circle P, with chord $\overline{BY}$ and intersecting chords $\overline{WB}$ and $\overline{YA}$, as shown below.

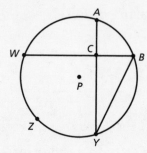

If $m\widehat{WZY} = 152°$, and $m\angle WCY = 84°$, what is the measure of $\angle AYB$?

A 16°

B 12°

C 8°

D 4°

CUMULATIVE EXAM

1. Which one of the following would be sufficient to conclude that $\triangle ABC \cong \triangle ZYX$?

   A $\angle A \cong \angle Z$, $\angle B \cong \angle Y$, and $\angle C \cong \angle X$.

   B $\angle C \cong \angle X$, $\overline{AC} \cong \overline{YZ}$, and $\angle B \cong \angle Y$.

   C $\overline{BC} \cong \overline{XY}$, $\angle B \cong \angle Y$, and $\overline{AB} \cong \overline{YZ}$.

   D $\overline{AB} \cong \overline{XY}$, $\angle C \cong \angle X$, and $\overline{BC} \cong \overline{XZ}$.

2. In quadrilateral *EFGH*, *EF* = 7 and *FG* = 10. Which one of the following could represent the lengths of the remaining two sides?

   A 4 and 19

   B 3 and 22

   C 10 and 30

   D 6 and 23

3. Consider circle *P*, as shown below.

   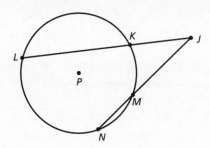

   If $m\widehat{LN} = 138°$ and $m\widehat{KM} = 74°$, what is the measure of $\angle J$?

   A 36°

   B 34°

   C 32°

   D 30°

4. *RSTU* is a parallelogram, and *WXYZ* is a non-isosceles trapezoid with bases of $\overline{WX}$ and $\overline{YZ}$. Which one of the following is <u>completely</u> correct?

   A $\angle R \cong \angle T$ and $\angle W \cong \angle Z$.

   B $\overline{RS} \cong \overline{TU}$ and $WX \neq YZ$.

   C $m\angle R + m\angle S = 90°$ and $\overline{WX}$ is parallel to $\overline{YZ}$.

   D $\overline{ST} \cong \overline{TU}$ and $m\angle Y + m\angle Z = 180°$.

5. Suppose the measures of three acute angles are added. Assuming that the measure of each angle is greater than 40° and less than 60°, what conclusion is true about their sum?

   A It is the measure of an obtuse angle.

   B It is the measure of an angle that is larger than an obtuse angle.

   C It is the measure of a right angle.

   D It is the measure of a straight angle.

6. △*ACE* ~ △*GJL*, as shown below.

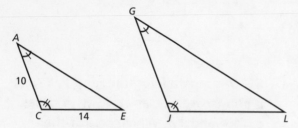

The ratio between corresponding sides from triangle *ACE* to triangle *GJL* is $\frac{2}{7}$. If *GL* is 55 units greater than *AE*, what is the perimeter of triangle *GJL*?

A   139 units

B   161 units

C   183 units

D   205 units

7. The longest side of the first of two similar triangles is 30, and its area is 216. The longest side of the second triangle is 25. What is the area of the second triangle?

A   128

B   144

C   150

D   180

8. Consider triangle *MNR* and its midsegments, as shown below.

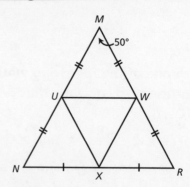

What type of triangle is represented by *UWX*?

A   Equilateral

B   Acute isosceles

C   Obtuse isosceles

D   Scalene

9. Which one of the following circumstances is <u>not</u> possible?

A   Two line segments that never intersect and are not parallel

B   A line that is parallel to a line segment

C   Two lines that never intersect

D   Two lines that are perpendicular bisectors of each other

10. Given two concentric circles, with circumferences of 36π and 16π respectively, what is the area of the annulus?

A   260π

B   180π

C   90π

D   10π

11. It is known that quadrilateral *BDFH*, which is neither a square nor a rectangle, is congruent to quadrilateral *KMRS*. Which one of the following <u>must</u> be true?

    A ∠*R* is not a right angle.

    B No two sides of *KMRS* are congruent.

    C If *m*∠*H* = 85°, then *m*∠*S* > 85 °.

    D If *BD* > *DF*, then *KM* > *MR*.

12. Consider circle *Q*, as shown below.

    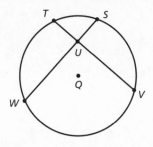

    If *TU* = 8 inches, *UW* = 18 inches, and *UV* is 9 inches larger than *SU*, what is the value of *UV*?

    A 7.2

    B 8.2

    C 15.2

    D 16.2

13. Consider the following diagram.

    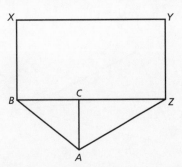

    How many of the angles located at vertices *X, Y, Z, A, B,* and *C* can be correctly named using only a single letter?

    A None

    B One

    C Two

    D Three

14. If a line intersects a circle twice, then this line must be a _____.

    A radius

    B secant

    C tangent

    D sector

15. In triangle *CDE, m*∠*E* = 56° and *m*∠*C* = 63°. Which one of the following inequalities is correct?

    A *CD* < *DE* < *CE*

    B *DE* < *CD* < *CE*

    C *CD* < *CE* < *DE*

    D *DE* < *CE* < *CD*

16. In triangle *FGH*, $\overline{GJ}$ bisects $\angle FGH$, as shown below.

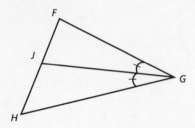

Which one of the following <u>must</u> be true?

A (*FG*)(*GH*) = (*FJ*)(*JH*)

B $\dfrac{FG}{GH} = \dfrac{FJ}{JH}$

C $\dfrac{FG}{GJ} = \dfrac{GJ}{GH}$

D The area of $\triangle FGJ$ equals the area of $\triangle GJH$.

17. $\triangle KLM \sim \triangle NPQ$, as shown below.

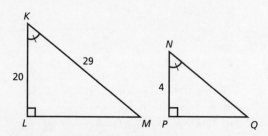

What is the value of *PQ*?

A 3.8

B 4.2

C 4.6

D 5.4

18. In circle *P*, shown below, $\overline{RV}$ is a tangent segment.

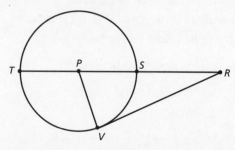

If *RT* = 24 and the radius of circle *P* is 5, what is the area of $\triangle PRV$ to the nearest hundredth?

A 45.83

B 47.85

C 55.87

D 57.89

19. $\triangle XYZ \sim \triangle BCD$, as shown below.

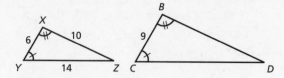

To the nearest hundredth, what is the area of $\triangle BCD$?

A 58.46

B 56.45

C 54.44

D 52.43

20. Which one of the following conditions will guarantee that rhombus *FHJL* is similar to rhombus *NPQR*?

A Their areas are equal.

B *FH* = (2)(*NP*)

C $m\angle H + m\angle Q = 180°$

D Their perimeters are equal.

21. In triangle *STV*, the altitude from point *V* is actually one of the sides. Which one of the following <u>must</u> be true?

    A The triangle is isosceles.

    B The triangle is obtuse.

    C The measure of ∠*V* is the smallest of the three angles.

    D The measure of ∠*S* or ∠*T* is 90°.

22. Consider the following diagram, in which points *X*, *Y*, and *Z* lie on line ℓ.

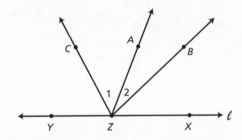

    If *m*∠*CZX* = 112°, *m* ∠*BZY* = 152°, and the measure of ∠1 is twice the measure of ∠2, what is the measure of ∠1?

    A 50°

    B 52°

    C 54°

    D 56°

23. Which one of the following triplets does <u>not</u> represent the sides of a right triangle?

    A 15, 36, 39

    B 18, 80, 82

    C 20, 25, 32

    D 24, 45, 51

24. In the diagram below, trapezoid *DFHJ* is similar to trapezoid *MPRT*.

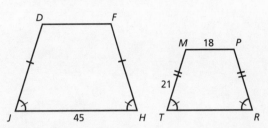

    If the ratio of the area of *DFHJ* to that of *MPRT* is $\frac{25}{9}$, what is the perimeter of trapezoid *MPRT*?

    A 87

    B 82

    C 76

    D 71

25. For quadrilateral *WXYZ*, it is known that its diagonals are perpendicular bisectors of each other. If it is <u>not</u> a square, which type of figure must *WXYZ* represent?

    A Rectangle

    B Trapezoid

    C Rhombus

    D Parallelogram

Answer Key

1

1. Two

An angle consists of two rays with a common endpoint.

2. Infinite number

It is not possible to identify each point of a line.

3. ∠A, ∠SAW, ∠WAS

Two ways in which an angle can be named are (a) by its vertex or (b) by three letters in which the middle letter is the vertex and each of the other two letters are points on a different ray.

4. ∠NSL or ∠LSN

Two adjacent angles share a common vertex and a common ray.

5. $\overline{AB}$, $\overline{BC}$, $\overline{AC}$

A line segment is named by using both endpoints. Note that *BA* is equivalent to *AB*. Likewise, the letters may be reversed for the other two answers.

6. B

A one-way street sign is shaped like an arrow.

7. D

Line segments are one-dimensional. They have length, but no width.

8. A

An angle may be described using only one letter if that letter is its vertex and there are no other angles at that vertex.

9. One

A point, which has no dimension, is named with just one letter.

10. ∠A, ∠PAM, ∠PAL, ∠MAP, ∠LAP

Two ways in which angle can be named are (a) by its vertex or (b) by three letters in which the middle letter is the vertex and each of the other two letters are points on a different ray.

Lessons

2

1. D   The relative size of segments is only applicable within a single problem.

2. A   Neither lines nor rays can be measured, since they extend indefinitely in at least one direction.

3. C   by definition of angle measurement

4. A   The measure of this angle is about 120°.

5. 8 $MJ = 21 - 13$

6. 48   Let x represent FH, so that $x + 11$ represents DF. Then $x + (x + 11) = 85$, $x = 37$, so $x + 11 = DF = 37 + 11$.

7. 14 $GK = 39 - 19 - 6$

8. 41 $TQ = 24 + 33 - 16$

9. 100°   Let x represent the measure of $\angle XVW$, so that $x + 62$ represents the measure of $\angle UVX$. Then $x + (x + 62) = 138$, $x = 38°$, so $x + 62 = m\angle UVX = 38° + 62°$.

10. 133°   Let x represent the measure of $\angle ZXY$, so that $7x$ represents the measure of $\angle BXC$. Then $x + 7x + 17 = 169$, which leads to $x = 19°$. Thus, $m\angle BXC = 169° - 19° - 17°$.

11. D $\angle ABC$ and $\angle CBD$ must be adjacent angles if $m\angle ABC + m\angle CBD = m\angle ABD$.

12. 118° $m\angle EFJ = 90 - 76 = 14°$. Then, $m\angle JFG = 132 - m\angle EFJ = 132° - 14°$.

3

1. D   The measure of an obtuse angle is greater than 90° but less than 180°.

2. C   The sum of the measures of two complementary angles is 90°.

3. B   The measure of each obtuse angle is greater than 90°, so their sum must be greater than 180°.

4. A   The sum of the measures of two supplementary angles is 180°.

5. B $\overline{AD}$ represents a line segment, $\overleftrightarrow{BF}$ represents a line, and the symbol $\perp$ means "perpendicular."

6. B   In a two-dimensional setting, two lines must either intersect or lie parallel to each other. The given lines are not parallel.

7. A   The symbol ‖ means "parallel."

8. D   In naming a 180° angle, the middle letter must be the vertex and the other two letters must lie on rays that go in opposite directions.

9. C   Since 190° is greater than 180°, it is neither an acute, a right, nor an obtuse angle.

10. B   The measure of $\angle TVX$ is about 85°.

Lessons

4

1. C   ∠7 is an obtuse angle, whereas ∠3 is an acute angle.

2. C   Alternate exterior angles lie on opposite sides of the transversal and also lie outside the parallel lines.

3. D   The measure of each of ∠3, ∠6, ∠8 is 30°

4. C   Corresponding angles lie in the same relative position with respect to the two parallel lines.

5. C   Let x represent the measure of ∠3, so that $x + 24$ represents the measure of ∠2. Then $x + (x + 24) = 180°$, $x = 78°$, and finally $m\angle 2 = 78 + 24$.

6. D   By the definition of "bisection," $BZ = ZF$. Statement II would only be true if BF bisected AC. Since these segments may intersect at any angle, Statement III is not necessarily true.

7. A   Based on the symbols that are embedded in the diagram, each of Statements I and II are correct. Although it appears that $ST = TQ$, they are not marked as such in the diagram.

8. A   Let x represent the measure of ∠4, so that $9x$ represents the measure of ∠6. Then $x + 9x = 180°$, and $x = 18°$.

9. B   Since lines do not have midpoints, they cannot be bisected.

10. A   One can draw an infinite number of lines that contain point M, although only one such line can be drawn that is perpendicular to $\overline{RS}$.

5

1. D   Each side of a polygon must be a line segment.

2. B   Beginning with any vertex, the polygon must be named by using letters associated with consecutive vertices. The letters C and K do not name consecutive vertices.

3. C   The perimeter is (7)(4).

4. C   The length of each side is 180 ÷ 5.

5. B   Let x and $3x$ represent the lengths of the other two sides. Then $50 + x + 3x = 122$ cm, so $x = 18$ cm. The largest side is represented by $3x$, which is (3)(18).

6. D   The sum of the lengths of any two sides of a triangle must exceed the length of the third side. $10 + 11 > 17$.

7. A   The sum of the lengths of any three sides must exceed the length of the fourth side. $3 + 5 + 7 < 16$.

8. B   A vertex always joins two sides.

9. C   Let x represent the length of the fourth side, so that $x + 9$ represents the length of the third side. Then $5 + 12 + x + (x + 9) = 42$, $x = 8$. Thus, the longest side is $8 + 9$ inches.

10. D   The sum of the lengths of any four sides must exceed the length of the fifth side. $5 + 5 + 8 + 10 > 27$.

Lessons

6

1. A   The measure of the third angle is 180 − 28 − 66 = 86°, so each angle has a measure less than 90°.

2. ∠N   The smallest angle must lie opposite the smallest side, which is $\overline{LM}$.

3. D $m\angle U = 180 - 40 - 35 = 105°$. Since $\angle U$ is the largest angle, the largest side must lie opposite it.

4. B   If two angles are each 60°, then the measure of the third angle must be 180 − 60 − 60 = 60°

5. A   In the same triangle, a side opposite a 50° angle would be smaller than a side opposite a 70° angle. Since *EF* > *IJ*, it must be true that △*DEF* is larger than △*HIJ*.

6. C   Since the two congruent angles each measure 41°, the third angle must measure 180 − 41 − 41, which is the measure of the largest angle.

7. C   Let *x* represent the measure of ∠*R*, so that 6*x* represents the measure of ∠*Q*. Then $x + 6x + 54 = 180$, $x = 18$, and finally $m\angle Q = (6)(18)$.

8. B   The third angle is 180 − 45 − 90 = 45°, so the triangle has a right angle and is also isosceles.

9. A   Each angle of an equilateral triangle is 60°, and 60 < 90.

10. D   Let *x* represent the measure of ∠*L*, so that *x* + 12 represents the measure of ∠*K*. Then $58 + x + (x + 12) = 180$. Solving, $x = m\angle L = 55°$ and $x + 12 = m\angle K = 67°$. Finally, 55 < 58 < 67.

7

1. remote interior

Definition of a remote interior angle. $\angle 1$ is an angle exterior to the triangle with vertex *R*. $\angle 3$ is an interior angle with vertex *Q*.

2. 137°

$m\angle 2 = 56° + 25° = 81°$. Then $m\angle 1 = 81° + 56°$.

3. B

By the given markings in the diagram, $\angle HJK \cong \angle KJI$.

4. C

When an altitude is drawn from an acute angle of an obtuse triangle, it lies completely outside the triangle.

5. 26°

Since the angle represented by the combination of $\angle 5$ and $\angle 6$ is an alternate interior angle to $\angle 4$, $116° = m\angle 5 + 90°$.

6. C

$m\angle 5 + m\angle 7 = 90°$, since $\angle 5$, $\angle 6$, and $\angle 7$ form a straight line, with $m\angle 6 = 90°$. Also note that $\angle 2 \cong \angle 5$, since they are alternate interior angles of parallel line segments. This means that $m\angle 2 + m\angle 7 = 90°$.

7. D

The medians of any triangle always intersect at an interior point of the triangle.

8. B

$\angle IKG$ is congruent to $\angle E$ because they are opposite angles of parallelogram *KIEG*. The only other two angles that are congruent to $\angle IKG$ are $\angle AIK$ and $\angle KGC$.

9. B

$AI = IE = KG = 6$.

10. $\angle KIG$, $\angle C$, and $\angle AKI$

$\angle KIG$ and $\angle IGE$ are alternate interior angles of parallel line segments, so they must be congruent. $\angle C$ and $\angle IGE$ are corresponding angles of parallel line segments, so they too must be congruent. Finally, $\angle AKI$ and $\angle C$ are corresponding angles of parallel line segments. By substitution, $\angle AKI \cong \angle IGE$.

Lessons

8

1. 12.5   Let x represent the length of $\overline{PR}$. Then $125 = \left(\frac{1}{2}\right)(x)(20)$, so $x = 125 \div 10$.

2. C   A median of a triangle joins a vertex to the midpoint of the opposite side. Using each half of the opposite side as a base, the corresponding height would be identical.

3. C and E   By the markings in the diagram, each of $\angle JKM$ and $\angle MKL$ has a measure of $\frac{1}{2} \times 90 = 45°$. By the Angle Bisector property, statement (E) is correct.

4. $33.\overline{3}$ The product of the base and corresponding height of a triangle is twice the triangle's area. Let x represent PR. Then, $36x = (30)(40)$, so $x = (30)(40) \div 36$.

5. A   The midsegments that join each of the midpoints of the triangle must be parallel to and one-half of the opposite side. However, no two midsegments need be congruent.

6. C $TU = 18$, so each of $\overline{UX}$ and $\overline{TX}$ must have a length of 9.

7. B and C   By definition, $\overline{YB}$ represents a median of $\triangle YZC$. For each of $\triangle YZB$ and $\triangle YCB$, YA represents the altitude. Since $ZB = BC$, the area of $\triangle YZB = \left(\frac{1}{2}\right)(ZB)(YA) = \left(\frac{1}{2}\right)(BC)(YA) =$ the area of $\triangle YCB$.

8. C   You will need to draw 3 medians, 3 angle bisectors, and only one altitude. The other two altitudes are already represented by the two legs.

9. B and E $\overline{DG}$ is an angle bisector, so $\angle EDG \cong \angle FDG$. $\overline{DG}$ is also an altitude, so $DG \perp EF$.

10. 58   Each of the four non-overlapping triangles has an area one-fourth that of $\triangle JKL$. Since the combined area of three of these triangles is 43.5, the area of each small triangle is $43.5 \div 3 = 14.5$. Thus the area of $\triangle JKL$ is $(4)(14.5)$.

1. 15.65   Let x represent YZ. Then, $x^2 = 14^2 + 7^2 = 245$.
$x = \sqrt{245}$.

2. 13.53   Let x represent VW. Then, $x^2 + 29^2 = 32^2$.
$x^2 = 183$, $x = \sqrt{183}$.

3. 7.59   Let x represent ST and let $3x$ represent RS.
Then $x^2 + (3x)^2 = 24^2$. So, $10x^2 = 576$, $x = \sqrt{57.6}$

4. 34.64   Let x represent NP and let $2x$ represent PQ.
Then, $x^2 + 30^2 = (2x)^2$. $3x^2 = 900$, $x = \sqrt{300} \approx 17.32$.
Finally, $PQ = (2)(17.32)$

5. 8   Let x represent DF. Then $x^2 + 15^2 = 17^2$. Finally, $x^2 = 64$, $x = \sqrt{64}$

6. 7.06   Let x represent DH. The product of base and its associated height is always equal to twice the triangle's area. Then $17x = (8)(15) = 120$.
So, $x = 120 \div 17$.

7. 9.8   Let x represent PL. Then $\frac{6}{x} = \frac{x}{16}$. So, $x^2 = 96$, $x = \sqrt{96}$.

8. 18.76   Let x represent LN. Using the value of PL from the previous exercise, $9.8^2 + 16^2 = x^2$. So, $x^2 = 352.04$, $x = \sqrt{352.04}$.

9. 80   Let x represent QM. Then $\frac{5}{20} = \frac{20}{x}$. So, $5x = 400$. Finally, $x = 400 \div 5$.

10. 188.08   Let x represent IK. Then $x^2 = 5^2 + 20^2 = 425$ $x = \sqrt{425} \approx 20.62$.
Now let x represent QK. Then $x^2 = 20^2 + 80^2 = 6800$.
$x = \sqrt{6800} \approx 82.46$. Perimeter of $\triangle IQK = 20.62 + 82.46 + 85$

Lessons

10

1. 32°   Let x represent the measure of $\angle D$ and let $x + 12$ represent the measure of $\angle C$. Then $x + (x + 12) + 168 + 116 = 360°$. So, $2x + 296 = 360$. Finally, $x = 64 \div 2$.

2. C   A figure with four congruent sides must be a square or a rhombus.

3. B   The diagonals of a rectangle are not perpendicular to each other.

4. 77   Let x represent width and let $7x$ represent the length. Then $(2)(x) + (2)(7x) = 176$. Then, $x = 176 \div 16 = 11$. The length is $(7)(11)$.

5. A   In a rhombus or a square, the diagonals bisect the angles that they intersect.

6. 10.47   The length of a diagonal is $(7.4)(\sqrt{2})$.

7. 22   Let x represent the length of the second diagonal. Then $\left(\dfrac{1}{2}\right)(9)(x) = 99$. So, $x = 99 \div 4.5$.

8. 65.48   Let x represent the length. $x^2 + 8^2 = 26^2$. $x = \sqrt{612} \approx 24.74$. Then the perimeter is $(2)(24.74) + (2)(8)$.

9. 960   The figure is a rhombus, so $RV = \dfrac{1}{2} \times 32 = 16$.

Let x represent UV and use the Pythagorean theorem in $\triangle RUV$. Then $x^2 + 16^2 = 34^2$. So, $x = \sqrt{900} = 30$. This means that $US = 60$.

The area is $\left(\dfrac{1}{2}\right)(32)(60)$.

10. C   A quadrilateral is named by using the letters for consecutive vertices. Vertices D and Y are opposite vertices.

Lessons

11

1. 13   Let x represent each of *CE* and *AG*. Since *EG* = 25,
             (2)(25) + 2x = 76. Then, x = 26 ÷ 2

2. 68°   Let x represent the measure of ∠*N*, which is congruent to ∠*J*. Let x +
             44 represent the measure of ∠*H*. Since ∠*H* and ∠*N* are consecutive
             angles, x + (x + 44) = 180. x = 136 ÷ 2

3. 424   Let x represent *SY*. So, $x^2 + 5^2 = 15^2$, $x = \sqrt{200} \approx 14.142$.
             Then the area is (30)(14.142)

4. 22.5   Let x and 3x represent the bases. Then $150 = \left(\frac{1}{2}\right)(10)(x + 3x)$.

             x = 30 ÷ 4 = 7.5, so the longer base is (3)(7.5).

5. B   Although the diagonals of an isosceles trapezoid are congruent, they
             do not necessarily intersect at a 90° angle.

6. 9.22   The area of a trapezoid is the product of its median and its height.
             Let x represent the height and let 5x represent the median.
             Then $(x)(5x) = 425$, $x^2 = 425 \div 5$, so $x = \sqrt{85}$.

7. 118   *RX* = *TV* = 27, since they represent the lengths of the non-parallel
             sides. Let x represent *VX*. Then 32 = (17 + x) ÷ 2, so x = 47.
             The perimeter is 27 + 17 + 27 + 47.

8. D   A rectangle is a type of parallelogram that has four right angles.

9. A   ∠3 and ∠4 are alternate interior angles of parallel lines, so that
             $m\angle 4 = 18°$. Also, $m\angle 2 = m\angle 1 + m\angle 4 = 42°$. The sum of the measures
             of ∠1 and ∠2 is 24° + 42°.

10. 342   *ZX* = (2)(*SZ*) = 30 and *ZT* = *ZY* + *YT* = 24. Let x represent *XT*.
             Then, $x^2 + 24^2 = 30^2$. Therefore, $x^2 = 324$, so x = 18.
             The area of *WXYZ* = (19)(18).

Lessons

12

1. B — $\overline{DK} \cong \overline{FJ}$, and $\overline{FJ}$ is the same line segment as $\overline{JF}$.

2. A — Regardless of whether $\angle H$ and $\angle M$ represent a pair of base angles or vertex angles, if $\angle H \cong \angle M$, each angle of $\triangle HLN$ can be matched with a corresponding angle of $\triangle MPR$. If one pair of sides is shown congruent, then the triangles are congruent by either Angle-Side-Angle or by Side-Angle-Angle.

3. Side-Angle-Angle — Based on the tick marks, there is a congruence for two pairs of angles and a pair of non-inclusive sides.

4. B — One angle in an obtuse triangle must exceed 90°. This size angle could never be found in an acute triangle.

5. B — Side-Side-Angle is not sufficient to guarantee a congruence between two triangles.

6. C — In any right triangle, the other two angles must be acute.

7. D — Answer choice (D) is correct because of Side-Angle-Side.

8. vertex, base — The vertex angle is formed by the two congruent sides. Base angles are formed by non-congruent sides.

9. C — Answer choice (C) would guarantee that the triangles are congruent by Side-Angle-Angle.

1. C $\overline{HF} \cong \overline{EJ}$, so we need to spot a congruence in which H is in the same position as E or J and also F is in the same position as E or J. Only answer choices (B) and (C) fulfill these requirements. However, in answer choice (B), $\overline{BH}$ is matched up with $\overline{CJ}$, which is impossible.

2. D   From the given information, the two figures must be rectangles. To guarantee that they are congruent, we must know that their respective lengths and widths are congruent.

3. B   Given any two squares, if any one corresponding pair of sides are congruent, then all four pairs of corresponding sides must be congruent. This implies that the squares are congruent.

4. C $\overline{WA}$ must be congruent to $\overline{BD}$.

5. C   The trapezoids must be congruent if all four pairs of sides and angles are congruent.

6. A   If $\overline{GJ} \cong \overline{ER}$, then all eight sides of the two rhombi have the same length. $\angle Q$ and $\angle V$ are supplementary to $\angle L$ and $\angle S$, respectively. Thus, if $\angle L \cong \angle S$, then $\angle Q \cong \angle V$. Since opposite angles of a rhombus are congruent, the remaining pairs of angles will be congruent.

7. D   The trapezoids are isosceles, so $\overline{AY} \cong \overline{HP}$ and $\overline{BZ} \cong \overline{IQ}$. Since $\overline{AY} \cong \overline{BZ}$, all four of these segments are congruent. Based on the tick marks, all pairs of corresponding sides are congruent. In addition, since $\angle Y \cong \angle Z$, all four lower base angles are congruent. $\angle A$ is supplementary to $\angle Y$, and this implies that all four upper base angles are also congruent.

8. A   If any one pair of corresponding angles are congruent, then each of the remaining pairs of corresponding angles can be shown to be congruent.

Lessons

14

1. C $(7)(0.4) \neq (77)(0.04)$

2. A Two pairs of sides are in proportion, plus the included angles are congruent.

3. D $\overline{JH}$, which is equivalent to $\overline{HJ}$, corresponds to $\overline{IK}$. Also, $\overline{LH}$ corresponds to $\overline{GK}$.

4. 17.5 Let x represent QS. $\dfrac{10}{7} = \dfrac{25}{x}$. Then, $x = (25)(7) \div 10$.

5. 21 Let x represent RQ and let $x + 9$ represent MP. Then, $\dfrac{10}{7} = \dfrac{x + 9}{x}$.

So, $10x = 7x + 63$. Finally, $x = 63 \div 3$

6. C Two right triangles may be similar without the condition that they are each isosceles. Thus, XW may not be equal to WY.

7. B If the triangles are similar, then $\dfrac{40}{32} = \dfrac{50}{GK}$.

This would imply that $GK = 40$, which equals the value of CE.

8. C Since $\angle N$ corresponds to $\angle M$, $m\angle M = 15°$.
Then, $m\angle Q = 180 - 115 - 15 = 50°$.

9. D If $\triangle STU \sim \triangle BFD$, then $\angle D \cong \angle U$ and $\angle F \cong \angle T$.
This would imply that $\angle D$ and $\angle F$ are base angles.

10. 32.73 $\dfrac{10}{15} = \dfrac{24}{HM}$, from which we can determine that $HM = 36$.

Let x represent KM. Then $x^2 + 15^2 = 36^2$. $x^2 = 1071$, so $x = \sqrt{1071}$.

Lessons

15

1. A $\overline{HK}$ corresponds to $\overline{ML}$, since both line segments lie between angles with one tick mark and four tick marks. Likewise, $\overline{KP}$ and $\overline{IL}$ are corresponding segments; they both lie between angles with three tick marks and four tick marks.

2. C Two squares are always similar.

3. D If any three angles of one quadrilateral are congruent, in pairs, to three angles of another quadrilateral, then the fourth pair of angles must be congruent. The two quadrilaterals will be similar, provided all pairs of sides are in proportion.

4. C $CX = DW = 12$. Also, since the rectangles are similar, $\frac{30}{12} = \frac{ZY}{6}$.

   Solving, $ZY = 180 \div 12 = 15$.

5. B In rhombus $EFGH$, $\angle E \cong \angle G$. In rhombus $VUTS$, $\angle V \cong \angle T$. If $m\angle E \neq m\angle T$, then $m\angle E \neq m\angle V$. Since these are corresponding vertices, the rhombi would not be similar.

6. A $\frac{9}{36} = \frac{16}{RQ}$, which yields $RQ = 64$. Then, $\frac{9}{36} = \frac{11}{PQ}$, which yields

   $PQ = 44$. So, $RQ + PQ = 64 + 44$.

7. B $XY = VZ$, so $\frac{24}{9.6} = \frac{18}{XY}$. Then, $XY = (18)(9.6) \div 24 = 7.2$

   Also, in an isosceles trapezoid, the measures of angles connected to one base are supplementary to the measures of the angles connected to the other base.

8. 70 $\frac{20}{XU} = \frac{28}{98}$. Then, $XU = (98)(20) \div 28 = 70$

Lessons

16

1. 12.5   Let x represent EF. Then, $\dfrac{2}{5} = \dfrac{5}{x}$. So, $x = 25 \div 2$.

2. B HJ and MK are corresponding sides. IJ and LK are also corresponding sides. Since $HJ > MK$, $IJ > LK$.

3. 14.62 $\dfrac{9}{6} = \dfrac{20}{RT}$, which yields $RT = 120 \div 9 = 13.\overline{3}$. Let x represent PT.

Then $x^2 + 6^2 = (13.\overline{3})^2$. So, $x^2 = 213.\overline{7}$. Finally, $x = \sqrt{213.\overline{7}}$.

4. 88 $ZY = 51 - 3 - 14 - 10 = 24$. Let x represent DB.

Then $\dfrac{3}{11} = \dfrac{24}{x}$, so $x = 264 \div 3$.

5. D   Suppose that the side of length 10 from $EFGH$ corresponds to the side of length 4 of $IJKL$. Also, suppose that the side of length 25 from $EFGH$ corresponds to the side of length 10 from $EFGH$.

Since $\dfrac{10}{4} = \dfrac{25}{10}$, these quadrilaterals would be similar.

6. 8.84 $\dfrac{5}{6} = \dfrac{MN}{30}$, so $MN = 150 \div 6 = 25$. Let x and $3x$ represent NP and MP,

respectively. Then $x^2 + (25)^2 = (3x)^2$. So, $8x^2 = 625$. Then $x = \sqrt{78.125}$.

7. 275   The perimeter of the first quadrilateral is 110.
Let x represent the perimeter of the second

quadrilateral. Then $\dfrac{40}{100} = \dfrac{110}{x}$, so $x = 11{,}000 \div 40$.

8. 30   Let x and $x + 9$ represent AB and CD, respectively. Then $\dfrac{7}{10} = \dfrac{x}{x + 9}$.

$7x + 63 = 10x$, so $x = 21$. Thus, $CD = 21 + 9$.

9. B, E   For any two squares, all angles are 90° and sides are in proportion. For any two equilateral triangles, all angles are 60° and sides are in proportion.

10. 112   Let x and $x + 21$ represent FM and EL, respectively. Then $\dfrac{x + 21}{x} = \dfrac{15}{8}$,

$15x = 8x + 168$. Solving, $x = 24$, which is the value of FM. Now let x

represent IM. Then $\dfrac{60}{x} = \dfrac{15}{8}$, so that $x = IM = (60)(8) \div 15 = 32$.

Finally, the perimeter of $FHIM$ is $(2)(24) + (2)(32)$.

Lessons

17

1. C   The ratio of corresponding sides must be $\sqrt{\dfrac{9}{25}} = \dfrac{3}{5}$. Note that $\dfrac{6}{10} = \dfrac{3}{5}$.

2. 210.94   Let x represent the area of $\triangle GIJ$. Using the reduced fraction of $\dfrac{8}{15}$ in place

of $\dfrac{40}{75}$, we can write $\left(\dfrac{8}{15}\right)^2 = \dfrac{64}{225} = \dfrac{60}{x}$. Then $x = (225)(60) \div 64$

3. 20.12   Let x represent the value of BH. Then $\left(\dfrac{27}{x}\right)^2 = \dfrac{9}{5}$.

So $\dfrac{729}{x^2} = \dfrac{9}{5}$, which means that $x^2 = (729)(5) \div 9 = 405$. Finally, $x = \sqrt{405}$.

4. B   The ratio of the areas is $\left(\dfrac{4}{8}\right)^2$.

5. 46.43   The semi-perimeter is $(8 + 15 + 21) \div 2 = 22$, so the area is $\sqrt{(22)(14)(7)(1)}$.

6. 124.84   Let x represent QR and QS. Then $2x + 25 = 57$, so $x = 16$. The semi-perimeter is $57 \div 2 = 28.5$, so the area is $\sqrt{(28.5)(12.5)(12.5)(3.5)}$.

7. D   The perimeter of $\triangle TUV$ is 36, so the ratio of the sides is $\dfrac{36}{30} = \dfrac{6}{5}$. The ratio of

the areas is $\left(\dfrac{6}{5}\right)^2$.

8. 2.90   The ratio of the sides is $\dfrac{8}{2} = \dfrac{4}{1}$. Then, $CJ = \left(\dfrac{1}{4}\right)(16) = 4$ and $HJ = \left(\dfrac{1}{4}\right)(12) = 3$.

In $\triangle CHJ$, the semi-perimeter is $(2 + 3 + 4) \div 2 = 4.5$,

so the area is $\sqrt{(4.5)(0.5)(1.5)(2.5)}$.

9. 33.67   Let x represent the largest side. Then $\left(\dfrac{18}{x}\right)^2 = \dfrac{20}{70} = \dfrac{2}{7}$. This means that

$\dfrac{324}{x^2} = \dfrac{2}{7}$, so $x^2 = (324)(7) \div 2 = 1134$. Finally, $x = \sqrt{1134}$.

10. 14.61   Let x represent KL, so that $x^2 + 30^2 = 34^2$. $x^2 = 256$, which means that

$KL = 16$. The area of $\triangle KLM$ is $\left(\dfrac{1}{2}\right)(16)(30) = 240$. The ratio of the areas

of these two triangles is $\dfrac{240}{200} = \dfrac{6}{5}$. Let x represent NP, so that

$\dfrac{6}{5} = \left(\dfrac{16}{x}\right)^2$. Then $\dfrac{6}{5} = \dfrac{256}{x^2}$, which means that $x^2 = (256)(5) \div 6 = 213.\overline{3}$.

Finally, $x = \sqrt{213.\overline{3}}$.

Lessons

18

1. D

Based on the lengths, the ratio of the areas should be $\left(\frac{44}{99}\right)^2 = \frac{16}{81}$. However, $\frac{200}{300} \neq \frac{16}{81}$.

2. 24.49

$\frac{16}{56} = \frac{2}{7}$. Let x represent the area of *MNPQ*. Then $\left(\frac{2}{7}\right)^2 = \frac{x}{300}$. So, $\frac{4}{49} = \frac{x}{300}$, which means that $x = (300)(4) \div 49$.

3. 29.02

The perimeter of Square #1 is 36 units. Let x represent the perimeter of Square #2. Then $\left(\frac{36}{x}\right)^2 = \frac{20}{13}$. So, $\frac{1296}{x^2} = \frac{20}{13}$, which means that $x^2 = (1296)(13) \div 20 = 842.4$. Finally, $x = \sqrt{842.4}$.

4. 35

Let x and $x + 14$ represent *RX* and *SY*, respectively. Then $\frac{25}{49} = \left(\frac{5}{7}\right)^2 = \left(\frac{x}{x + 14}\right)^2$. We can simplify this proportion to $\frac{5}{7} = \frac{x}{x + 14}$. $7x = 5x + 70$, so $x = 70 \div 2$.

5. 42.60

$\frac{840}{150} = \frac{28}{5}$. Let x represent *FB*. Then $\frac{28}{5} = \left(\frac{x}{18}\right)^2$. Thus, $\frac{28}{5} = \frac{x^2}{324}$, which means that $x^2 = (324)(28) \div 5 = 1814.4$, so $x = \sqrt{1814.4}$.

6. 39.44

The area of a rhombus is one-half the product of its diagonals. Let x represent *ZD*. Then $840 = \left(\frac{1}{2}\right)(42.60)(x) = 21.30x$, so $x = 840 \div 21.30$.

7. 11.23

Let x represent *LP*. Then $2x + (2)(59) = 189$. Solving, $x = (189 - 118) \div 2 = 35.5$. Since *QT* corresponds to *LP*, $\frac{10}{1} = \left(\frac{35.5}{x}\right)^2$. Thus, $\frac{10}{1} = \frac{1260.25}{x^2}$, which means that $x^2 = 1260.25 \div 10 = 126.025$, so $x = \sqrt{126.025}$.

8. B

Two rhombi are similar if all corresponding angles are congruent. If each of two corresponding angles has a measure of 80°, then each corresponding consecutive angle has a measure of 100°. Consequently, all four corresponding angles are congruent.

Lessons

9. 34   Let x represent CU. Then $\frac{144}{200} = \frac{18}{25} = \left(\frac{10}{x}\right)^2$. Thus, $\frac{18}{25} = \frac{100}{x^2}$, which

means $x^2 = (100)(25) \div 18 = 138.\overline{8}$, so $x = \sqrt{138.\overline{8}} \approx 11.79$. Applying

the area formula to trapezoid $CWDX$, $200 = \left(\frac{1}{2}\right)(11.79)(CW + XD)$.

Finally, $CW + XD = 200 \div 5.90$, rounded off to the nearest integer.

10. 15.73   The area of $EFGH$ is $\left(\frac{1}{2}\right)(6)(18 + 21.5) = 118.5$. The area of $QRST$ is

118.5 − 28 = 90.5. Let x represent QR. Then $\frac{118.5}{90.5} = \left(\frac{18}{x}\right)^2$.

Thus $\frac{118.5}{90.5} = \frac{324}{x^2}$, which means that $x^2 = (324)(90.5) \div 118.5 \approx 247.44$.

Finally, $x = \sqrt{247.44}$.

19

1. A   A reflex angle measures between 180° and 360°.

2. C   The degree measure of an arc is equivalent to the degree measure of its associated central angle.

3. C   By definition, a segment represents the area bounded by a chord and its associated minor arc.

4. C   A chord is a line segment joining any two points on a circle. The diameter is the longest chord.

5. ∠1, ∠3 $\overparen{AE}$ is the intercepted arc for each of ∠1 and ∠3, since each of A and E are the endpoints of the chords comprising ∠1 and ∠3.

6. 52° $m\angle UPW = 360 - 98 - 210$.

7. 50° $195 = m\overparen{QRS} + m\overparen{RST} - m\overparen{RS}$. Thus, $m\overparen{RS} = 112 + 133 - 195$.

8. B   Two circles are internally tangent if they intersect at one point, and one of them lies completely inside the other.

9. A, E   Two concentric circles share the same center. However, their radii are different. This implies that their circumferences are also different.

10. 65° $m\angle GPH + m\angle HPK +$ measure of reflex ∠$GPK = 360°$. The degree measure of $\overparen{HK} = m\angle HPK = 360° - 70° - 225°$.

11. Inscribed   Definition of a triangle inscribed in a circle.

Lessons

20

1. 17.19   The diameter equals $54 \div \pi$.

2. C $\pi R^2 = 500\pi$, which means that the radius equals $\sqrt{500} \approx 22.36$. Then the circumference equals $(2)(\pi)(22.36)$.

3. 4π The circumference is $(2)(\pi)(9) = 18\pi$.

The linear measure of $\overset{\frown}{CE}$ is $\left(\dfrac{80}{360}\right)(18\pi)$.

4. 315°   The circumference is $(2)(\pi)(16) = 32\pi$.

The degree measure of reflex $\angle GQJ$ is $\left(\dfrac{28\pi}{32\pi}\right)(360°)$.

5. D   The area of the circle is $(\pi)(30^2) = 900\pi$.

The area of the sector is $\left(\dfrac{135}{360}\right)(900\pi)$.

6. 975π The area of the annulus is $(\pi)(40^2) - (\pi)(25^2)$.

7. A   Let x represent QW. $56\pi = (\pi)(12^2) - (\pi)(x^2)$, so $x^2 = 12^2 - 56 = 88$. Then $x = \sqrt{88} \approx 9.38$. Finally, $VW = (2)(9.38)$.

8. $36\pi - 72$ The area of the segment equals the area of the sector minus the area of the triangle $= \left(\dfrac{1}{4}\right)(\pi)(12^2) - \left(\dfrac{1}{2}\right)(12)(12)$.

9. B   The area of the sector is equals $\left(\dfrac{40}{360}\right)(\pi)(32^2) = 113.\overline{7}\pi$. The semi-perimeter of the triangle is $(32 + 32 + 22) \div 2 = 43$, so its area is $\sqrt{(43)(11)(11)(21)} \approx 330.55$ The area of the shaded segment is approximately $113.\overline{7}\,\pi - 330.55$.

10. D   The central angle is $\left(\dfrac{31\pi}{81\pi}\right)(360°)$.

21

1. $\angle B$ Each of $\angle A$ and $\angle B$ are inscribed angles and their measures are one-half of $m\widehat{DC}$.

2. 153°   Let x represent the measure of $\angle QFH$. Since $m\angle QFH = m\angle QHF$, $x + x + 54 = 180$. Then $2x = 126$, so $x = 63°$. A tangent ray is always perpendicular to a radius at the point of intersection, so $m\angle HFG = 90 + 63$.

3. 72°   Let x and $4x$ represent the measures of $\widehat{IJ}$ and $\widehat{JK}$, respectively. Since $m\widehat{IJK} = 180°$, $x + 4x = 180°$. Then $5x = 180°$, so $36°$. This means that $m\widehat{JK} = 144°$. Finally, $m\angle I = \left(\dfrac{1}{2}\right)(144)°$.

4. 186° $m\widehat{MN} = (2)(44) = 88°$. Let x and $x + 100$ represent the measures of $\widehat{LN}$ and $\widehat{MRL}$, respectively. Then $x + (x + 100) + 88 = 360°$. $2x = 360 - 100 - 88 = 172°$, so $x = 86°$. Thus, $m\widehat{MRL} = 100 + 86$.

5. Two   A right angle must exist at $\angle T$ and at $\angle U$, since a tangent ray forms a right angle with a radius at the point of intersection.

6. 129°   Let x represent $m\widehat{WZY}$. Then $26 = \left(\dfrac{1}{2}\right)(x - 77)$. Multiplying by 2, this equation becomes $52 = x - 77$. Thus, $x = 77 + 52$.

7. 26° $m\widehat{DA} = (2)(m\angle C) = 124°$. Thus, $m\widehat{AB} = 150 - 124$.

8. 65° $m\angle DEC = \left(\dfrac{1}{2}\right)(26 + 104) = \dfrac{1}{2} \times 130°$

9. 107° $m\widehat{FHJ} = 360 - 70 - 76 = 214°$. So, $m\angle L = \dfrac{1}{2} \times 214°$

10. 202°   The sum of opposite angles of an inscribed polygon is 180°. Let x and $x + 22$ represent the measures of $\angle F$ and $\angle J$, respectively. Then $x + (x + 22) = 180°$. The next steps are $2x = 158°$, $x = 79°$, $m\angle J = 101°$, and finally $m\widehat{LFH} = (2)(101)$.

Lessons

22

1. C — All radii of a circle are equal and two tangents drawn to a circle from the same external point are equal.

2. 18 — Let x represent EQ. $x^2 + 16^2 = 30^2$. Then $x^2 = 16^2 + 30^2 = 1156$, which means that $x = 34$. $EF = EQ - QF = 34 - 16$.

3. 5.62 — Let x represent PG. $x^2 = 10^2 + 12^2 = 244$. Then $x = \sqrt{244} \approx 15.62$ So, $JG = PG - PJ = 15.62 - 10$.

4. C — $(KN)^2 = (KM)(KL) = (32)(13) = 416$. So $KN = \sqrt{416}$.

5. 14 — Let x represent TS, so that $32 - x$ represents RS. $(RU)^2 = (RT)(RS)$, so $24^2 = (32)(32 - x)$. Then $576 = 1024 - 32x$, which means that $x = 448 \div 32$.

6. 4.74 — Let x and $4x$ represent VY and VZ, respectively. $(VX)(VW) = (VZ)(VY)$, so $(18)(5) = (4x)(x)$. $90 = 4x^2$, which means that $x = \sqrt{\dfrac{90}{4}} = \sqrt{22.5}$.

7. A — $(AE)(AC) = (AG)(AJ)$. Let x represent AJ. Then $(28)(8) = 25x$, so $x = 224 \div 25$.

8. 8.5 — Let x represent WT. $(LW)(WT) = (NW)(WR)$. Then $2.5x = (3)(5)$, so $x = 15 \div 2.5 = 6$. Finally, the diameter, LT, is $6 + 2.5$.

9. 6.32 — Let x represent each of BJ and JD. $(FJ)(JH) = (BJ)(JD)$. Then $(2.5)(16) = x^2$. Thus, $x = \sqrt{40}$.

10. 12.5 — Let x represent MZ. $(KZ)(MZ) = (SZ)(VZ)$. Then $8x = (5)(20)$. Thus, $x = 100 \div 8$.

1

1. B   For any two line segments, one of the following conditions will occur: (a) they intersect, (b) they are parallel to each other, or (c) they neither intersect nor are they parallel to each other.

2. C   If $m\angle 2 = 90°$, then $m\angle 1 = 180 - 90 = 90°$

3. D $FG = 40 - 15 - 12$

4. B   A ray goes in one direction and has one endpoint.

5. D   Acute angles are those whose measure is less than 90°.

6. A   The box at the intersection of $\overline{QM}$ and $\overline{KP}$ shows that the segments are perpendicular to each other. The different number of tick marks indicates that neither segment bisects the other.

7. A   Alternate exterior angles lie outside the two parallel lines and on opposite sides of the transversal.

8. B   Let x and $x + 12$ represent ST and RS. Then, $x + (x + 12) = 84$. $2x = 72$, so $x = 36$. Finally, $RS = 36 + 12$.

9. C   An angle is named using a point on one ray (or segment), followed by the vertex, followed by a point on the other ray (or segment).

10. D $m\angle QKN = m\angle QKM - m\angle NKM = 25°$. Then, $m\angle QKL = m\angle QKN + m\angle NKL = 25 + 90$.

Quizzes

2

1. C   The measure of the third angle is 180° − 50° − 31° = 99°, which is an obtuse angle.

2. A   Let x represent CE. Then, $x^2 + 7^2 = 18^2$. Then, $x^2 = 275$, so $x = \sqrt{275}$.

3. C   Let x and $x - 15$ represent the measures of $\angle J$ and $\angle L$, respectively. Then $x° + (x - 15)° + 45° = 180°$. So, $2x = 180° + 15° - 45° = 150°$. Consequently, $x = 75°$ and $x - 15 = 60°$. This means that $m\angle G < m\angle L < m\angle J$. The sides opposite these angles must be in the same order of inequality.

4. D   Let x and $4x$ represent fourth and third sides, respectively. Then $10 + 16 + x + 4x = 51$. So, $5x = 25$, which means $x = 5$ and $4x = 20$. The two largest sides are 16 inches and 20 inches.

5. C $\angle V \cong \angle PTR$, since they are corresponding angles of parallel lines. Also, $\angle XPT \cong \angle PTR$, since they are alternate interior angles of parallel lines. Thus, $\angle V \cong \angle XPT$.

6. A   In a triangle, the product of any base and the altitude drawn to that base is twice the area. Thus, $(24)(DF) = (30)(18)$, which means that $DF = 540 \div 24$.

7. B   A polygon must be named by using consecutive letters, either by moving clockwise or counterclockwise.

8. D $m\angle 6 = m\angle 4 = 48°$ and $m\angle 7 = 180° - m\angle 1 - m\angle 6 = 69°$.

9. C   Let x represent XZ. Then, $ZY = 27 - 9 = 18$ and $\dfrac{WZ}{XZ} = \dfrac{XZ}{ZY}$.
By substitution, $\dfrac{9}{x} = \dfrac{x}{18}$. So, $x^2 = 162$. Finally, $x = \sqrt{162}$.

10. B   Let x and $x + 8$ represent GL and LH, respectively.
Then $\dfrac{BG}{BH} = \dfrac{GL}{LH}$. By substitution, $\dfrac{16}{26} = \dfrac{x}{x + 8}$.
This leads to $26x = 16x + 128$, so $x = 128 \div 10 = 12.8$.
Finally, $LH = 12.8 + 8$.

3

1. A   A Side-Side-Angle correspondence does not guarantee congruence between two triangles.

2. A   A congruence between one pair of lengths and one pair of widths establishes a congruence between two rectangles.

3. C   Angle-Side-Angle means a congruence between two pairs of corresponding angles and a congruence between the included sides.

4. D   Let x and $3x$ represent FJ and HG, respectively. Then $(3x)(x) = 261$, so $x^2 = 261 \div 3 = 87$. Thus, $x = \sqrt{87} \approx 9.33$ which means that $HG = (3)(9.33)$

5. A   The diagonals of a rhombus are perpendicular bisectors of each other. $PZ = ZT = 16$, so $PT = 32$. Also, $PR = \left(\frac{1}{4}\right)(84) = 21$. Let x represent ZR in $\triangle PRZ$. Then $x^2 + 16^2 = 21^2$, which leads to $x^2 = 185$. So, $x = \sqrt{185} \approx 13.6$, which means that $KR = (2)(13.6) = 27.2$ Thus, $PT - KR = 32 - 27.2$.

6. B   The area is $\left(\frac{1}{2}\right)(13)(34 + 16) = \frac{1}{2} \times 650$ square inches.

7. D   Since each leg of the first isosceles right triangle is larger than each leg of the second isosceles right triangle, the hypotenuse of the first triangle must be larger than the hypotenuse of the second triangle.

8. C   Let x represent the length of each diagonal. Then $46.24 = \left(\frac{1}{2}\right)(x)(x)$, which means that $x^2 = 92.48$. Thus, $x = \sqrt{92.48}$ inches.

9. B   The median must be eight units shorter than the longer base, which implies that the median is 40. The area of a trapezoid is the product of its median and its height. Thus, its height is $500 \div 40$.

10. D   Based on the given tick marks, $ZY = DC$. Since $\angle A \cong \angle W$, it follows that $\angle X \cong \angle B$. Each of the four angles $\angle Z$, $\angle Y$, $\angle D$, $\angle C$ is supplementary to any of the angles $\angle W$, $\angle X$, $\angle A$, $\angle B$. Thus, $\angle Z \cong \angle Y \cong \angle D \cong \angle C$.

Quizzes

4

1. B Let x represent the perimeter of the larger triangle.

Then $\sqrt{\dfrac{100}{169}} = \dfrac{10}{13} = \dfrac{15}{x}$. So, $10x = 195$. Finally, $x = 195 \div 10$.

2. C Let x represent GH. Then $\dfrac{8}{20} = \dfrac{10}{x}$. $x = 200 \div 8 = 25$.

Now let x represent JK. Then $\dfrac{8}{20} = \dfrac{16}{x}$.

So, $x = 320 \div 8 = 40$. Finally, $JK - GH = 40 - 25$.

3. A Since the two rectangles are not similar, $\dfrac{LM}{QR} \neq \dfrac{MN}{RS}$.

Note that $\dfrac{18}{26} \neq \dfrac{8}{12}$.

4. C Let x represent YZ. Then, $\dfrac{18}{12} = \dfrac{27}{x}$. Then, $x = 324 \div 18 = 18$.

Now let x represent XZ. Then $x^2 = 12^2 + 18^2 = 468$. $x = \sqrt{468}$.

5. A Let x represent CL. Then $\dfrac{4}{9} = \dfrac{8}{x}$. So, $x = 72 \div 4 = 18$.

Now let x and $x + 25$ represent GJ and HL, respectively.

Then $\dfrac{4}{9} = \dfrac{x}{x + 25}$. So, $9x = 4x + 100$.

$x = 100 \div 5 = 20$, so that $HL = 45$.
The perimeter of $CFHL$ is $(2)(18) + (2)(45)$.

6. A $\dfrac{16}{4} = \dfrac{10}{2.5}$ and the included angles are congruent.

The triangles are similar by Side-Angle-Side.

7. D The semi-perimeter is $(18 + 11 + 15) \div 2 = 22$.
The area is $\sqrt{(22)(4)(11)(7)} = \sqrt{6776}$.

8. B Let x represent UV. The ratio of the areas is $\dfrac{250}{175}$, which reduces to $\dfrac{10}{7}$.

Then $\dfrac{10}{7} = \dfrac{x^2}{9^2}$. So, $7x^2 = 810$, which means that $x \approx \sqrt{115.71}$.

9. D Two rhombi are similar, but not congruent, if all pairs of corresponding angles are congruent but the sides are not congruent. If $m\angle F = 110°$, then $m\angle C = 180° - 110° = 70°$. Since $m\angle C = m\angle G$, this is sufficient to conclude that the given rhombi are similar.

10. C Let x and $x + 900$ represent the areas of $QRST$ and $LMNP$, respectively.

The ratio of corresponding sides is $\dfrac{42}{28} = \dfrac{3}{2}$. Then $\dfrac{x + 900}{x} = \left(\dfrac{3}{2}\right)^2 = \dfrac{9}{4}$.

So $9x = 4x + 3600$, which means that $x = 3600 \div 5 = 720$. Thus, the area of $LMNP$ is $720 + 900$.

Quizzes

5

1. B $m\angle J = 180° - 84° = 96°$, since opposite angles of an inscribed quadrilateral must be supplementary. Then, $m\widehat{BEG} = (2)(96°)$.

2. D   The circumference is 18π. Let x represent the measure of $\angle MQN$. Then $\dfrac{x}{360} = \dfrac{7\pi}{18\pi}$, so $x = 2520\pi \div 18\pi$.

3. B   A chord is a line segment connecting any two points on a circle.

4. A   Let x represent the measure of $\widehat{TZ}$. Then, $m\angle R = \left(\dfrac{1}{2}\right)(\widehat{VX} - \widehat{TZ})$, so by substitution, $32° = \left(\dfrac{1}{2}\right)(102° - x)$. Then $64° = 102° - x$, followed by $x = 38°$. Finally, $m\widehat{TVZ} = 360° - 38°$.

5. C $m\widehat{CHE} = m\widehat{CG} + m\widehat{EH} - m\widehat{HG}$. Let x represent $m\widehat{GH}$. Then $180° = 133° + 109° - x$. Thus, $x = 242° - 180°$.

6. A   Let x and $x + 3$ represent LN and KL, respectively. Then $(8)(x) = (6)(x + 3)$, which simplifies to $8x = 6x + 18$. So, $x = 18 \div 2 = 9$. Finally, $NR = 9 + 8$.

7. A   By definition, concentric circles have the same center but different radii. They cannot intersect.

8. D   The radius of this circle is $60\pi \div 2\pi = 30$ so the circle's area is 900π. Let x represent the area of the sector. Then, $\dfrac{x}{900\pi} = \dfrac{150}{360} = \dfrac{5}{12}$. Thus, $x = 4500\pi \div 12$.

9. B   Since the radius is 10, $VU = 20$. Then, $SU = SV - VU = 16$. Let x represent ST. $(ST)^2 = (SV)(SU)$, so $x^2 = (36)(16) = 576$. Thus, $x = \sqrt{576}$.

10. C   Let x represent $m\widehat{AB}$. Then $m\angle WCY = \left(\dfrac{1}{2}\right)(\widehat{WZY} + \widehat{AB})$. $84° = \left(\dfrac{1}{2}\right)(152° + x)$. So, $168° = 152° + x$, followed by $x = 16°$. Finally, $m\angle AYB = \dfrac{1}{2} \times 16°$.

Cumulative Exam

1. C   The triangles would be congruent by Side-Angle-Side. (Congruencies of Triangles)

2. A   The sum of any three sides must exceed the fourth side. $7 + 4 + 10 > 19$. (Perimeters of Polygons)

3. C $m\angle J = \left(\frac{1}{2}\right)(m\widehat{LN} - m\widehat{KM})$.

Thus, $m\angle J = \left(\frac{1}{2}\right)(138° - 74°) = \frac{1}{2} \times 64°$.
(Angle formed by Two Secants)

4. B   Opposite sides of a parallelogram are congruent and the bases of a trapezoid are not congruent. (Properties of Parallelograms and Trapezoids)

5. A   The sum of the three angles would be between 120° and 180°. (Definition of Obtuse Angle)

6. B   Let x and $x + 55$ represent AE and GL, respectively. Then $\frac{2}{7} = \frac{x}{x + 55}$.
$7x = 2x + 110$, so $x = 110 \div 5 = 22$. This means that $GL = 77$.
$\frac{2}{7} = \frac{10}{GJ}$, so $GJ = 35$. Finally, $\frac{2}{7} = \frac{14}{JL}$, so $JL = 49$.
The required perimeter is $77 + 35 + 49$. (Similar Triangles)

7. C   Let x represent the area of the second triangle.
The ratio of corresponding sides is $\frac{30}{25} = \frac{6}{5}$. Then $\frac{216}{x} = \left(\frac{6}{5}\right)^2 = \frac{36}{25}$.
Thus, $x = 5400 \div 36$. (Similar Triangles)

8. B $UX = MW = MU = WX$, since a midsegment is half the length of the opposite side. The measure of each of $\angle N$ and $\angle R$ is $(180° - 50°) \div 2 = 65°$. Since $\angle UXW \cong \angle M$, $\angle UWX \cong \angle N$, and $\angle WUX \cong \angle R$, triangle UWX also contains angles with measures of 50°, 65°, and 65°. (Properties of Midsegments)

Cumulative Exam

9. D   A line cannot be bisected, since its length is infinite. (Properties of Lines)

10. A   The radii of the two circles are 18 and 8, respectively. The area of the annulus is $(\pi)(18^2) - (\pi)(8^2) = 324\pi - 64\pi$. (Areas of Circles)

11. D   Given that $BD > DF$, simply substitute KM for BD and substitute MR for DF. (Congruent Triangles)

12. D   Let x and $x + 9$ represent SU and UV, respectively. Then $(x)(18) = (x + 9)(8)$, which simplifies to $18x = 8x + 72$. This means that $x = 72 \div 10 = 7.2$. Then UV = 7.2 + 9. (Intersecting Chords in a Circle)

13. C   Only at points X and Y can an angle be named using just one letter. (Properties of Angles)

14. B   A secant is a line that intersects a circle twice. (Properties of Secants and Tangents)

15. C $m\angle D = 180° - 63° - 56° = 61°$. Within one triangle, the largest side is opposite the largest angle. Likewise for the smallest side. In this example, $m\angle E < m\angle D < m\angle C$. (Relationship Between Angles and Sides of a Triangle)

16. B   An angle bisector of a triangle divides the two segments that comprise the angle into the same ratio as the two portions of the opposite side. (Angle Bisector Theorem)

17. B $LM^2 + 20^2 = 29^2$, which simplifies to $LM^2 = 441$.

So, $LM = \sqrt{441} = 21$. Let x represent PQ. Then $\dfrac{20}{4} = \dfrac{21}{x}$.

Thus, $x = 84 \div 20$. (Similar Triangles)

18. A   Since $PT = 5$, $PR = RT - PT = 19$. Let x represent RV. Then $x^2 + 5^2 = 19^2$, which simplifies to $x^2 = 336$.

So, $x = \sqrt{336} \approx 18.33$. Since $m\angle V = 90°$, the area of

$\triangle PRV$ is approximately $\left(\dfrac{1}{2}\right)(18.33)(5) = (9.165)(5)$.

(Pythagorean Theorem and Tangents to a Circle)

Cumulative Exam

19. A   The semi-perimeter of $\triangle XYZ$ is 15, so its area is determined by $\sqrt{(15)(9)(5)(1)} \approx 25.98$. The ratio of corresponding sides is $\frac{6}{9} = \frac{2}{3}$. Let x represent the area of $\triangle BCD$. Then $\frac{25.98}{x} = \left(\frac{2}{3}\right)^2 = \frac{4}{9}$. So, $x = 233.82 \div 4$. (Similar Triangles and Heron's Formula)

20. C   In rhombus $FHJL$, $m\angle H + m\angle J = 180°$. To assure similarity between two rhombi, corresponding angles must be congruent. If $m\angle H + m\angle Q = 180°$, then $\angle J \cong \angle Q$. Since $\angle J$ and $\angle Q$ are corresponding angles, the rhombi are similar. (Similar Polygons)

21. D   Either $\overline{VS} \perp \overline{ST}$ or $\overline{VT} \perp \overline{ST}$. In the first instance, $m\angle S = 90°$; in the second instance, $m\angle T = 90°$. (Definition of Altitudes)

22. D $180° = m\angle BZY + m\angle CZX - m\angle CZB$. Let x represent $m\angle CZB$. By substitution, $180° = 152° + 112° - x$, so $x = 84°$. Now let x and $2x$ represent the measures of $\angle 2$ and $\angle 1$, respectively. Then, $x + 2x = 3x = 84°$, so $x = 28°$. Finally, $m\angle 1 = (2)(28°)$. (Adjacent Angles)

23. C   In a right triangle, the sum of the squares of the two legs must equal the square of the hypotenuse. $20^2 + 25^2 = 1025 \neq 32^2$. (Pythagorean Theorem)

24. A $PR = MT = 21$. Let x represent RT. The ratio of the corresponding sides is $\sqrt{\frac{25}{9}} = \frac{5}{3}$, so $\frac{45}{x} = \frac{5}{3}$. This means that $x = 135 \div 5 = 27$. The perimeter of $MPRT$ is $21 + 18 + 21 + 27$. (Similar Polygons)

25. C   For a rhombus and a square, the diagonals are perpendicular bisectors of each other. (Properties of Quadrilaterals)

Workspace

Workspace

Workspace

Workspace

Workspace

Workspace

Workspace

Workspace

Workspace

Workspace

SCORECARD

Geometry

Lesson	Completed	Number of Drill Questions	Number Correct	What I need to review...
1		10		
2		12		
3		10		
4		10		
5		10		
6		10		
7		10		
8		10		
9		10		
10		10		
11		10		
12		10		
13		8		
14		10		
15		8		
16		10		
17		10		
18		10		
19		11		
20		10		
21		10		
22		10		

Quiz		What I need to review...
1	/10	
2	/10	
3	/10	
4	/10	
5	/10	

Cumulative Exam	/25	